Building Materials Evaluation Handbook

Forrest Wilson

VAN NOSTRAND REINHOLD COMPANY
NEW YORK CINCINNATI TORONTO LONDON MELBOURNE

Manufactured in the United States of America

Published by Van Nostrand Reinhold Company Inc.
135 West 50th Street,
New York, New York 10020

Van Nostrand Reinhold Company Limited
Molly Millars Lane
Wokingham, Berkshire RG11 2PY, England

Van Nostrand Reinhold
480 Latrobe Street
Melbourne, Victoria 3000, Australia

Macmillan of Canada
Division of Gage Publishing Limited
164 Commander Boulevard
Agincourt, Ontario MIS 3C7, Canada

15 14 13 12 11 10 9 8 7 6 5 4 3 2 1

Library of Congress Cataloging in Publication Data

Wilson, Forrest, 1918-
    Building materials evaluation handbook.

    Includes bibliographical references and index.
    1. Building materials—Testing—Handbooks,
manuals, etc.   2. Building failures--Handbooks, manuals,
etc.   I. Title.
TA410.W49 1983   691   83-6833
ISBN 0-442-29325-9

*for Pat Grado and Joe Marzeki,
architects and friends*

How to Use This Book

This is a reference book. Although it might conceivably be read in the order in which the subjects appear it was designed to be consulted subject to subject as one uses a dictionary or encyclodpedia. To facilitate quick identification and location of building materials, characteristics and problems they are first listed in the table of contents, repeated in the chapter headings and listed in the index.

In addition to describing how building materials respond to environmental stresses in terms of their mechanical, electrical, chemical and thermal properties, brief references to their normal behavior and a comparison of various material charcteristics has been included.

Most of the information gathered and presented here represents the contemporary developments of ancient building lore. The increasing importance of renewal, rehabilitation, retrofit and restoration is placing added importance on material behavior. A separate and distinct field of building science is emerging as increasingly sophisticated instruments are linked to the growing ability and decreasing costs of computer analysis. This book describes one segment of a new building science—that of building diagnosis.

Acknowledgments

This is to acknowledge and thank the following people and organizations for their assistance and generous sharing of knowledge and kind permission to quote from their writings and publications.

Mr. Harold Olin, AIA and John L. Schmidt, AIA and Walter H. Lewis, AIA, and the U.S. League of Savings Associations for permission to quote from their excellent and valuable book, *Construction: Principles, Materials and Methods*.

The Brick Institute of America (BIA) for permission to quote from their 'Technical Notes.'

Ms. Anne Grimmer, Preservation Assistance Division, National Park Service of Washington D.C. and all of the people that have prepared the "Preservation Briefs." Some of those that have contributed to these excellently researched and popularly written documents are listed here; Robert C. Mack, AIA, de Teel Patterson Tiller, James S. Askins, Baird M. Smith, AIA, Sarah M, Sweetser, John H. Meyers to name but a few of those under the technical editorship of Lee H. Nelson, AIA. The Technical Preservation Services Division, Heritage Conservation and Recreation Service, U.S. Department of the Interior, Washington D.C.

James H. Pielert and Porter Driscoll of the National Bureau of Standards Center for Building Technology, for their generous help, assistance and council.

Mr. Elmer Botsai, FAIA, Dean of the School of Architecture of the University of Hawaii at Manoa. Elmer was a building "pathologist," long before this activity was made popular by the conservation and restoration movements. He generously allowed the inclusion of exerpts from his forthcoming book on moisture infiltration and permitted me to publish an interview with him. Dean Botsai is a man of forthright opinions. He is sometimes easy to disagree with, but impossible to dislike.

Mr. Neal Fitzsimons of the Engineering Council, Kensington Maryland. Neal is a civil engineer and historian who has done and continues to do pioneering work in the field of building distress. My gratitude for advice, help, friendship and permission to quote from his study, "Research Support for Building Rehabilitation; Studies in the Area of Strength and Stability Evaluation."

Progressive Architecture, for permission to quote from its October 1966 article on precast concrete panels, November 1977 article, "Receipts for Baked Earth," and February 1981 article, "Your Solution Or Your Leak."

Canadian Building Digest.

Technology and Conservation Magazine.

National Research Council of Canada and its Division of Building Research for permission to quote from "Cracks, Movements and Joints in Buildings", NRCC 15477.

American Society for Metals, for permission to quote from "Nondestructive Inspection and Quality Control, Metals Handbook, Volume 11, 8th Edition.

The Concrete Reinforcing Steel Institute for historic information on reinforcing steel systems in old reinforced concrete structures.

Perhaps the single most important reference in this book is to the work of Frank H. Lerchen, James H. Pielert and Thomas K. Faison for their preparation of "Selected Methods for Condition Assessment of Structural, HVAC, Plumbing and Electrical Systems in Existing Buildings, NBSIR 80-2171, Center for Building Technology, National Engineering Laboratory, National Bureau of Standards.

These are but a few of the dedications to the people that supplied the information for this book. I was merely their scribe. You will find their names and those of many more reappearing throughout this book. I have tried to credit those quoted immediately after their work. For this reason the bibliography at the end of the book is brief. Authors and their works are scattered throughout the book as close as possible to those parts of the book to which reference to them is made.

Contents

Tower of Brooklyn Bridge, New York City, 1875. (Photo by F. Wilson)

INTRODUCTION

Three sleek new skyscrapers inch their way to completion in a five-block stretch of Madison Avenue, a tribute to the continuing popularity of midtown Manhanttan as one of the world's most elegant and expensive addresses.

Beneath the skyscrapers—which will house such blue-chip tenants as American Telephone and Telegraph, International Business Machines, and Continental Illinois National Bank—the city's water and sewer system decays.

Limousines clog the Wall Street area each day, whisking the captains of business to their appointed rounds, and each night the chauffeur-driven cars line up at Le Cirque, Regine's and the Plaza.

But the drivers take their passengers down the FDR Drive at some risk, for the major East Side highway is crumbling. The landfill underneath is slipping into the East River, and concrete chunks regularly break off from the ceilings of the drive's tunnels. . . .

In addition, New York must find another $20 billion to $30 billion to rebuild the rest of its physical plant. It must replace much of its 2,400 mile water and 6,100 mile sewer system (much of it is more than 100 years old). The city must repair its bridges; the Manhattan Bridge can sway several feet when a subway crosses, and cables snap on the Brooklyn Bridge (one killed a pedestrian in the summer of 1981). It must repave a large portion of more than 6,000 miles of streets.

"The outlook is grim," former deputy mayor Solomon says of the city's problems. "But that may be true everywhere, not just here."

—*The Washington Post,* Business and Finance F1, April 4, 1982

The deterioration of city services and city buildings dramatically, sometimes tragically, emphasizes two major environmental problems. One, economic necessity to fully utilize existing buildings has grown far beyond the fashionable historic preservation movement, which was originally based on the desire to conserve a national architectural heritage. The second problem is payment now demanded for past neglect of the built environment. The fact that buildings stand up at all is something of an accomplishment. To presume they will stand by themselves, impervious to time and the elements, is foolhardy. Humans seek to make their buildings outdo nature, for all natural materials break down in a progressive series of disintegrations due to chemical, physical, and biological activities. All around us mountains wear down, trees fall, exposed iron pyrites corrode, natural acids convert one element into another, roots split stone, and men and women die. Time changes the properties of materials. The designer uses his or her skill to control the rate of change by limiting and controlling the forces

Photo by F. Wilson

of deterioration. Once a building has been built a constant battle must be waged against destructive natural forces of disintegration.

People were once well aware of the transitory nature of life and buildings, but somehow, shut up in an increasingly urban environment, we have been lulled into believing that mechanical, chemical, thermal, and electrical forces are totally controlled. It just isn't so—instead, they have grown more destructive.

There is a growing national concern, born of necessity, for the full utilization of buildings that have been built. Buildings that have stood the test of time are now realized to be social anchors that stabilize neighborhoods. Familiar architectural configurations help unify and preserve neighborhood fabrics, devastated by their demolition and the construction of new buildings.

Subway entrance, Grand Central Station, New York City. (Photo by F. Wilson)

Cable connection Brooklyn Bridge, New York City. (Photo by F. Wilson)

Terrazzo pavers patched with plywood, New York City Financial District. (Photo by F. Wilson)

Economic factors cannot be overlooked in the growth of the building rehabilitation industry. Increasing costs of materials, labor, land acquisition, utilities, and financing have forced building costs to far exceed the general increase in the cost of living. The housing industry, which in the best of times does not provide sufficient housing to meet the populations needs, is made further ineffective.

The Department of Commerce reported that 42.2 billion dollars was spent in 1979 on "maintenance, repair and construction improvements to residential properties," compared to 77.1 billion spent in the construction of new housing units. The magazine *Architectural Record* reported that expenditures for nonresidential additions, alterations, and major replacement were expected to increase from an estimated 15 billion in 1978 to as much as 30 billion annually by the mid 1980s. A recent F. W. Dodge report stated that 77% of all construction activity in 1981 involved preservation, adaptive reuse, and renovation.

Architects and builders are faced with new problems in renewal, reuse, alteration, and the preservation of buildings. Another concept of the use of materials and methods of architectural construction is demanded, calling on different skills and techniques. Added to the problem of making a building stand up are those of how well and for how long it will do so. Building diagnosis has become an important field of building science.

The art of building must now share importance with the art of preserving buildings. The latter seeks out and finds the root causes of building deterioration and decay and then ingeniously seeks to solve these problems. This activity requires a high level of analytical skill, observation, and creative imagination which is certainly equal to the initial skill in the design and building of buildings.

Structural engineers provide the necessary information and prescribe the necessary materials to resist and counteract external building forces and the continuing tendency of inanimate matter to disintegrate. Designers must design sufficient resistances into their buildings to withstand gravitational pull and lateral forces as well as internal disintegration stresses. Chemists must detect the malignant decaying properties conveyed by air and deposited on building façades waiting for the catalyst of moisture to awaken their destructive malignant appetites. Any omission will result in failure of designed performance.

Yet despite the best intentions of engineer, architect, and builder, failures can and do occur. They take place more often than we care to admit. Most failures have similar shapes and therefore probably similar causes. A record of building difficulties and causes of distress is of value in reducing future incidence of failure.

Failures are quite democratic. They occur in all types of structures, framed or wall bearing, timber, steel, concrete, plastic, marble, or cream cheese. Foundation failures are so common, says Jacob Feld, Consulting Engineer, that if we define failure as the noncompliance of the structure with design expectations a structure that behaves as anticipated can be hailed as an "engineering triumph."

As structures have become statically more complicated they have compounded the possibility and incidence of secondary stresses induced by loadings other than traditional, vertically supported dead and live loads, and there has been an accompanying greater incidence of problems.

It is simply very difficult to anticipate all the conditions and combination of elements, natural and man made, that act upon a building. The best test is always the actual conditions themselves. In spite of increasing ability and analysis and greater depths of knowledge on the part of designers and builders, the proof of a successful building remains that it stands over time.

A lack of general understanding of the difficulties of building, the variety of forces, and the chemical action of building materials often results in a continued misuse of materials and combinations of materials.

Unfortunately, leaking composite walls, aerodynamic separation of wall and roof covering from the parent structure, delayed shrinkage of certain lightweight-aggregate masonary blocks and concrete mixes, the crystallization and brittle failure of some metal alloys, and electrochemical disintegration of certain metal imbedments in concrete with planned and unplanned admixtures and climatic exposure are failures not deterred by fame, Feld reminds us. They occur on and in the structures of the most prestigious architects as well as those of the common run.

Similar histories of difficulties can be observed in all building materials and all phases of the building industry. They are not limited to architectural construction, but occur in steel bridges, simple engineered buildings, tanks and silos, timber frames and roofs, and footings and foundations of all types and of all details.

A history of building difficulties is invariably a history of building materials and those who design and build buildings. When Portland cement as we know it was devised, a safe and economical construction material came into being. But this event is comparatively recent in building history. The standardization of Portland cement did not take place until after the first decade of this century. The reinforcing steel used today also evolved through a number of less effective prototypes and was standardized at about the same time as Portland cement. Many buildings standing today that engineers, architects, and historians are endeavoring to save were made with cement rock and antiquated structural steel reinforcing. The lessons of past failures, Feld reminds us are the foundation for the rational design of reinforced concrete as we know it today. Incidentally no one assumes that concrete failures or the design of concrete has come to an end.

When incidents of failure become too prominent, especially if loss of life occurs, governmental regulations are enacted. Legal codes appear that limit the

Cables and tower of Brooklyn Bridge, New York City. One of these cables
snapped and killed a pedestrian on this walkway. (Photo by F. Wilson)

freedom of thought and action of those involved in the design and construction
of buildings.

There is also a new phenomenon in the history of architecture that designers
and diagnosticians must contend with. It is a creature of the industrial revolution
called *technical building obsolescence*. In times past buildings were altered and
adapted as uses changed. As medieval cities moved beyond their walls some
walls were demolished but others remained in the city and became houses or
parts of buildings, as can be seen today in cities such as Florence, Rome, and
Moscow. Since the coming of the industrial revolution, buildings have been
designed and built for more specialized functions. As the ability to calculate
structures more exactly advanced, buildings were designed increasingly for single, specialized uses. Redundancy of structure, which allowed for adaptive reuse,

gave way to optimization. But buildings designed for a single function can become both technically and structurally obsolete. And so, buildings became transitory. They were torn down and new ones built almost seasonally. The witticism heard in New York City in the 1950s and 1960s—"It's a great city if they ever get it finished"—was directed at the phenomenon of constant rebuilding. This was characteristic of cities worldwide. The one consistent urban form, recognizable in all the cities the world over, were the huge cranes hovering over building skeletons. The residents of Honolulu dubbed these their "national bird" in the boom years of construction.

During this period of nervous building and destruction, the renewal, retrofit, or preservation of buildings was considered appropriate for only a few isolated historic structures as curtain wall skyscrapers were built and dismantled around them.

Preservation was, for a numbers of years, a narrow architectural concentration pursued by eccentric architects, ladies' clubs, art historians, and interested dilettantes. These times have changed, it has now become a vital field of American building. Buildings that during the 1950s and 1960s would never have been thought worthy of preservation are now considered valuable.

But as buildings last longer the forces that can destroy them have more time to work. The environment has become increasingly malignant. Noise and vibration abound, airborne acids coat buildings, and other environmental conditions not fully understood act on them to their detriment. Steel, concrete, and glass

Manhattan seen from Brooklyn Bridge. (Photo by F. Wilson)

have not been in use as long as brick, stone, and wood. We do not have enough time to construct a history of their responses, and the environment they and we inhabit today is totally unique compared to what we have experienced in the past. But along with the change in the building environment and the alarming degeneration of materials has come increased engineering ability. For as the skill to calculate startling new structures has been developed, so has a comprehensive ability to evaluate old ones.

This book is an introduction to the exciting new art, science and detective work of building material diagnosis. It is the study of the art of building through the study of building difficulties. Its practitioners are imaginitive, engineers, architects, art historians, interested lay men and women and students who are concerned and fascinated by the life and life expectancies of the buildings around them.

REFERENCES

Feld, Jacob, *Failure Lessons in Concrete Construction: Articles Collected from Concrete Construction Magazine,* 1978.

U.S. Dept. of Commerce, U.S. Bureau of the Census, *Annual Housing Survey: 1977,* Sept. 1979.

U.S. Dept of Commerce Construction Reports, *Residential Alterations and Repairs, Annual 1979,* issued April 1980.

Kidd, Phillip E., "Value of Non-Residential Rehabilitation Will Double by Mid-1980's," *Architectural Record* Oct. 1979, p. 61.

BUILDING DIAGNOSIS—Destructive and Nondestructive Testing

Rehabilitating a building rather than building a new one presents unique construction problems. Most of the building systems within it are affected. The examiner is involved not only in how the systems were originally designed and their interrelationships, but also in how they have survived in relation to each other. It is important to assess the condition of the entire system to determine the ability of its individual parts to support modifications.

Investigating techniques vary from simple visual inspections to complex laboratory and in-situ testing which may require extremely complex equipment operated by skilled, trained operators.

Essentially there are two methods. The first is nondestructive evaluation, which is the process of inspecting, evaluating, and measuring the properties of materials or systems without changing, damaging, or destroying their properties or affecting the service life of the test specimen. The second is destructive evaluation, which is the process of inspecting, evaluating, and measuring the properties of materials or systems in a manner which can change, damage, or destroy the properties or affect the service life of the test specimen.

Assessing the condition of a building is but one phase of an overall rehabilitation process which includes social, economic, political, and technical factors. It is only the technical factors bearing on building materials that concern us here.

The preliminary phase of building investigation is based on information collected in both on- and off-site examinations. It is quite helpful to know the date of construction of a structure so that the investigator can use the then current codes and journals as a fairly reliable source of design practices and strengths of materials. In cases where conclusive documentation is not available, dates can be estimated by architectural style.

Finding and studying good documentary evidence of the design and or construction phase of a building is the most satisfactory method to evaluate its behavior. If documents are not available, more costly and time-consuming methods, or even field and laboratory tests, may be the only way to obtain sufficient data on which to base opinions. The best solution, of course, is to obtain original design and construction drawings and contracts and to have the option of using field and laboratory tests to confirm the contract documents.

One of the keys to success in the evaluation of materials and systems of building is an ability to recognize the existing conditions which require evaluation and then to use testing techniques best suited to assess these conditions.

Fortunately, alternatives such as those suggested by structural engineer Neal Fitzsimons and listed below are available:

1. Review tests previously conducted and interpolate the effect elapsed time may have on estimates of strength properties.
2. Field test certain structural components using laboratory tests where needed and analytically apply the test results to evaluate the entire structure.
3. Test scaled-down models with either the same or analogous materials.
4. Analyze the performance of similar structures and apply the results of the analysis to the structure in question.

Current methods of analyzing the properties of materials and systems can be very helpful in determining safe load capacities and performance levels as well as limit states (deflection, stability, and the like). But the soundness of any evaluation still rests ultimately on experience and sound judgment on the part of the investigator Fitzsimons warns. It should be remembered that, while the evaluation methods described here are generally recognized as good practice it is advisable to consult local codes and building officials on any specific local performance requirements for the building and its systems.

REFERENCES

Fitzsimons, Neal, C.E., *Research Support for Rebuilding Rehabilitation Studies in the Area of Strength and Stability Evaluation, National Bureau of Standards, 1979.*
Lerchen, Pielert, and Faison, *Selected Methods for Condition Assessment of Structural, HVAC, Plumbing and Electrical Systems in Existing Buildings,* U.S. Department of Commerce, National Bureau of Standards, Wash. D.C., 1980.

I. The Nature of Materials

Diagnosis

The Building Blocks of Matter

Compounds

Molecules

Bonding of Matter

Ionic Bonding
Metallic Bonding
Covalent Bonding
Secondary Bonding

The Properties of Materials

Mechanical Properties

Stress and Strain
Elastic Deformation
Plastic Deformation
Stress Strain Test

Hardness

Thermal Properties

Melting Temperatures
Thermal Conductivity
Thermal Expansion
Electrical Properties
Chemical Properties

Galvanic Series of Common Metals and Alloys

DIAGNOSIS

The medical profession uses diagnosis as a valuable tool. They have spent centuries gathering and organizing information about human response to maladies. For the physician diagnosis is not an end in itself. It is linked to the prediction of future events. *Prognosis* is the next step after diagnosis.

Medical knowledge includes a history of diseases. Time is an important dimension of these histories. The doctor anticipates the series of events and forms a prognosis based on his knowledge of the disease measured in time. The quality, precision and application of this knowledge will determine the outcome of events for the patient.

Prognosis is not an end in itself any more than diagnosis. The physician must decide if the future it predicts is acceptable, and choose to intervene. This intervention is termed *therapy*.

When the doctor initiates therapy he or she anticipates that with the passage of time the patient who might have arrived at an undesirable state will instead be in better condition. If therapy is successful the disease is deflected from its natural course.

But therapy is not always undertaken. In some cases it is not available and in others the prognosis may be benign. In the latter instance the disease can be allowed to run its course unimpeded. If the disease and its risks are well known the dangers of intervention can be intelligently weighed. A high probability of a dangerous result of untreated disease may justify a therapy that carries a high risk of injury.

Good diagnosis is based on good models of the condition diagnosed. Normal states must be defined so that abnormal states can be recognized.

In medicine diagnostic evidence is separated into *symptoms* and *signs*. Symptoms are told to the doctor by the patient. Signs are objective physical evidences of abnormal states and include measurements made with instruments.

Much of medical diagnosis is based upon pattern recognition, which is the ability to organize data into coherent structures. Humans do this with astonishing skill. Computers in comparison, at the present time, are programmed with only primitive recognition capabilities. It is therefore necessary for human diagnosticians to train themselves to recognize patterns of building disease, although computers can supply them with a great wealth of information to aid in visualizing these patterns.

To begin, the building diagnostician must have a memory model in his mind of the normal state or condition to be diagnosed. We therefore begin with the composition of building materials. We will then discuss the nature of each material and its response to mechanical, chemical, electrical, and thermal stresses as an aid to building a memory model to aid in diagnosis, prognosis, and eventual therapy.

THE BUILDING BLOCKS OF MATTER

There are about 100 chemical substances which cannot be subdivided. These basic units, such as carbon, iron, hydrogen, and oxygen are called *chemical elements*. Oxygen and silicon make up more than three-fourths of all the earth's matter, and just eight chemical elements constitute more than 97 percent of all earthly substances.

One of the great wonders of the world is that its magnificent complexity is formed from such a limited number of ingredients.

A given quantity of any element can be divided into smaller, smaller, and yet smaller quantities, until finally only a minute particle remains. This final entity, the smallest particle, which retains the properties of the original material from which it came, is called the atom.

Atoms are the basis of all matter. They consist of a small, dense, positively charged nucleus surrounded by a moving ring or rings of negatively charged electrons. The structure of the atom has been compared to that of the earth's solar system, with the electrons moving around the nucleus like our planets move around the sun.

Ordinarily electrons and nucleus are nicely balanced, the negative charged on the electrons equalling the positive charged on the nucleus. However, atoms may either give up or acquire negatively charged electrons, and are then termed *ions*. And they may combine with other atoms into larger units called *molecules*. All matter is formed from these three basic building blocks: atoms, ions, and molecules.

Substances are chemically classified and described according to their properties, including degree of solubility in various solvents, hardness, color, melting and boiling points, crystal form, taste, odor, density, and combustibility.

Compounds

These are substances composed of two or more elements in chemical combination whose proportions, by weight, are definite and invariable. A compound is homogenous, that is each molecule of it is like any other molecule. For example, water is a compound of hydrogen and oxygen; its molecule contains two atoms of hydrogen in chemical combination with one of oxygen. A compound has properties peculiar to itself, which are distinct from the properties of its constituents. For example, table salt, or sodium chloride is composed of the metal sodium and the greenish, poisonous gas chlorine, yet it does not resemble either of its constituents. In general, compounds can be decomposed by heat, pressure, or chemical reaction with certain other elements or compounds.

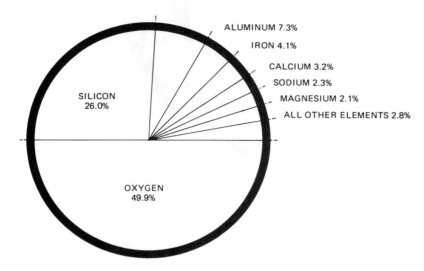

A handful of elements make up the bulk (97%) of all earthly matter. (After Construction: Principles, Materials and Methods—U.S. League of Savings Association)

Protons and neutrons are much denser than electrons, so essentially all of the weight of an atom resides in the nucleus. The sum of the weights of the protons and neutrons of one atom of an element is called the *atomic weight* of that element. Since the atom is electrically neutral, the positive charge of the protons in the nucleus is balanced by an equal negative charge from the electrons, so that the number of electrons is equal to the number of protons; this is the *atomic number* of the element.

Groups of electrons moving in neighboring orbits are said to belong to the same shell. The first shell, closest to the nucleus, can only accommodate 2 or 3 electrons, the second a maximum of 8, and the third 18 and the fourth 32. The distribution of electrons around the nucleus gives elements some of their inherent chemical properties.

Under certain conditions electrons may leave an atom, upsetting the balance and changing it into a positive ion. This tendency is common in elements having few electrons in their outermost shells, since the positive attraction of the nucleus for these remote electrons is weak. This condition is characteristic of metallic elements.

Other atoms have a tendency to acquire excess electrons in their outer orbits and become negative ions. These elements, usually with five or more electrons in their outermost shells, are almost invariably nonmetallic.

Molecules

A real material may consist of billions of one or more kinds of atoms (atomic substances), or billions of several kinds of ions (ionic substances). Still other materials are formed by the joining together of molecular units, each made up of

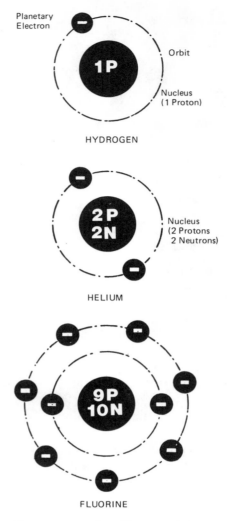

1P

Orbit

Nucleus
(1 Proton)

HYDROGEN

2 P
2N

Nucleus
(2 Protons
2 Neutrons)

HELIUM

9P
10N

FLUORINE

Electrons move in orbits around the nucleus. Helium's two electrons move in the
same orbit, but fluorine's 9 electrons move
in several different orbits. Groups of electrons with identical or neighboring orbits
comprise a shell; hydrogen and helium have
one shell, fluorine two. Quoted from *Construction, Principles, Materials and Methods* by permission of the U.S. League of
Savings Associations.

a number of atoms. A molecule of methane, for example, much like the water molecule described earlier, is composed of one carbon atom and four hydrogen atoms. Such molecules can only be broken into their constituent atoms with great difficulty.

According to the kinetic molecular theory of matter, molecules are in constant vibratory motion. The state or form in which matter appears—solid, liquid, or gas—depends upon the distance between the molecules and upon their velocity. In general, in a solid the molecules are relatively close together, the cohesive forces between them is strong, and their velocity is relatively slow; in a liquid they are further apart and move at a faster rate, but their motion is restricted within a definite volume by cohesive forces between the molecules. In a gas they are furthest apart and move at greatest velocity, the attractive force between them being small. The kinetic energy, that is the energy of motion of the molecules, is heat energy. Therefore when heat is added to a substance its molecules generally begin to vibrate rapidly and move apart. If the pressure upon the substance is kept constant, the substance expands. When heat is withdrawn, the activity of the molecules decreases, they come closer together, and the substance contracts. If heat is added continuously to a solid or a liquid, the molecules of the substance are stimulated to move at such velocities and at such distances from each other that a change of state occurs, from solid to liquid or from liquid to gas.

Some believe that in the absence of all heat the motion of the molecules would cease entirely. Pressure as well as temperature affects the molecular condition of a substance and therefore its state. For example, when pressure of sufficient magnitude is applied to a gas, the molecules of the gas are forced more closely together, their velocity is decreased, heat is given off, and the gas changed to a liquid.

Molecules exist in a great variety of shapes and sizes. The three atoms of carbon dioxide lie in a straight line, whereas those of water form an isosceles triangle. Methane is a tetrahedron of hydrogen atoms with a carbon atom at its center. More complex molecules form rings (benzene), chains (sugar), and helices (DNA) and (RNA), among other shapes. Like or unlike molecules may combine to form polymers (polymerization); typical polymers are the chain molecules of rubber and of many plastics, long strings of hydrocarbon molecules connected end-to-end. Substances differ according to the structure and composition of their molecules. Molecules of different substances differ in size and weight, as well as structure. In chemical action the structure of the molecules of the substance taking part is affected, the atoms breaking from their original molecular arrangement and recombining to form other molecules and therefore different substances.

Substances composed of molecules are called *molecular materials*. Water is a molecular material. When large numbers of the triangular water molecules are bonded closely together they exist as ice. As the tightness of the bond is loosened water is liquid. When bonding is very loose, it exists as water vapor. In all three forms the triangular units are intact, and the individual hydrogen and oxygen atoms do not leave their respective molecules.

Molecules can vary in size. A molecule of water has only 3 atoms but a molecule of methane has 8 and butane 14.

Bonding of Matter

One of the major reasons that atoms, ions, and molecules bond together to form substances is the strong acquisitive nature of the outermost electron shell. When a shell has the maximum number of electron pairs it is filled and occupies a low energy condition on the atomic scale, it is relatively inert.

All systems in nature strive to remain in a state of lowest energy. A partially filled outer shell has energy to spare, causing its atom to be chemically reactive. The filled-outer-shell condition produces materials which are stable and resist change. This filled outer shell is in general achieved by chemical bonding. (Elements with filled outer shells are inert gases, helium, neon, argon, krypton, and radon. These refuse to combine with any other elements, including each other, except under the most extreme laboratory conditions.)

There are four distinct types of bonding: ionic, metallic, covalent, and secondary. Only the last of these departs from the "outer shell" norm.

Ionic Bonding

When atoms with almost empty outer electron shells are juxtaposed with atoms with almost filled shells their natural tendency is to seek a state of stability. The atoms with few outer electrons surrender their "free" electrons and become positive ions, whereas atoms with many outer electrons accept these electrons to fill their outer orbit and become negative ions. Positive ions and negative ions are now in close proximity. They are attracted and bond ionically. Metals tend to have almost empty outer orbits, and nonmetals tend to have almost filled outer shells.

The most familiar ionically bonded substance is common table salt. Atoms of sodium have a single outer electron and therefore have a strong tendency to form positive ions. Atoms of chlorine have seven outer electrons and upon annexing one more fill their outer orbits and achieve stability as negative ions.

Ceramic materials are combinations of metallic and nonmetallic atoms bonded, primarily, through the ionic mechanism. Examples of this category of materials are brick, tile, Portland cement (which forms concrete), and natural stone.

Their properties follow from the quality of their ionic bond. They have high melting temperatures and are chemically inert. They tend to be brittle because the regular and extremely rigid arrangement of positive and negative ions resists change in shape: they tend to shatter, rather than to change shape. Brittleness is generally associated with good strength in compression but low strength in tension. These materials are also poor conductors of heat and electricity, because their electrons are tightly bound into the stable ions and are not free to conduct electrical or thermal energy in response to applied voltage or temperature.

It is apparent that the typical properties of ceramic materials follow logically from the nature of their constituent ions and of the ionic bond. Hundreds of new ceramic materials may be developed in the next few years but all of them will have high melting temperatures, high compressive strength, chemical inertness, and good thermal and electrical resistance.

Metallic Bonding

Metallic atoms try to surrender their few outer electrons to become positive ions. When large numbers of metallic atoms come together they each contribute elec-

Terra cotta decoration, U.S. Pension Building, Washington, D.C. (Photo by F. Wilson)

trons to a mobile "sea" which circulates near the original parent atoms. The metal ions are now in a stable situation existing in a three-dimensional metal crystal, with an electron "sea" dispersed throughout the solid.

The characteristic properties of metals follow from the nature of this bond. They are strong, with fairly high melting temperatures. Unlike ceramics, they are good conductors of heat and electricity because the electrons in the "sea" are extremely mobile and free to transport thermal and electrical energy.

Metals may or may not be chemically inert. Their ability to transport electrical current makes them susceptible to chemical degradation in the form of corrosion.

Cast iron drain pipe cover. (Photo by F. Wilson)

The corrosion tendency sometimes outweighs the intrinsic strength of the metallic bond.

Covalent Bonding

Many elements such as carbon and nitrogen lack the strong tendency to form either positive ions or negative ions. These elements with a moderate number of electrons in their outer shells reach stability by sharing electrons with similar elements. The process of mutual sharing of outer valence electrons by a cluster of atoms to create a stable entity is known as covalent bonding.

Covalent bonding is the most common mechanism whereby small numbers of atoms are bound into molecules. These molecules are then joined together by weak secondary bonds, to produce molecular materials. The covalent bond itself is very strong. It is very difficult to break up a molecule into its constituent atoms.

A few substances are produced entirely by covalent bonding. Diamonds are a familiar example. These substances are extremely hard and have high melting temperatures due to the strength of the covalent bonds. However, they seldom occur in nature. Diamonds are rarely found, and then only in small quantities.

Molecular materials are composed of atoms bound into molecules by covalent bonding, but the molecules are then joined to each other by means of weak secondary bonds. These bonds occur as the positive nuclei or negative electrons in one molecule feel an attraction of their opposites in neighboring molecules and are attracted to and bound to them in a weak bond.

Molecular materials, e.g., wood, plastics, and bituminous products, derive their properties from the behavior of this secondary bond. They have low strength and low melting temperatures compared to metals and ceramics, because the weak secondary bond is easily altered by the application of heat or force. These

Rotting tree. (Photo by F. Wilson)

materials are poor conductors of heat and electricity because individual electrons remain tightly constrained within covalently bonded molecules.

Surprisingly this weak secondary bond is not broken by many of the strong chemical compounds that attack metals and ceramics. Thus molecular materials are often chemically inert in a large number of environments. They are attacked by molecular solvents such as acetone, but are resistant to attack by most salts, acids, and industrial atmospheres.

THE PROPERTIES OF MATERIALS

The sharp differences in properties among materials allow us to divide building materials into three major categories: (1) The ceramics and glasses, brick, and concrete, are hard, brittle, and poor conductors of heat and electricity. (2) The metals—iron, steel, copper, aluminum, and alloys that do wondrous things with strange names and numbers—are more ductile than ceramics and are good conductors of heat and electricity. (3) In "molecular" materials the molecules themselves are not as firmly held together as the metallic atoms. These materials have fair strength and low melting temperatures, and are poor conductors of heat and electricity. The most common of these in building are wood and plastics.

To investigate how a material has performed and why it disintegrates or fails under stress begins by first identifying the category in which it belongs and then examining the properties of that category in relation to how it was strained by the stresses placed upon it by the building's geometry. What are its mechanical, electrical, thermal and chemical properties and how did these fare in the stressful real world?

Construction site, New York City. (Photo by F. Wilson)

Although building materials don't talk they do express themselves. As anyone knows who has heard the creaks and groans of a building during the quiet of the night, its materials are always working. They respond to external stimuli, heat, wind, static and dynamic loads. We can use all of our senses to examine them. They can be heard, seen, smelled, felt, and even tasted, although *taste* usually refers to intangible qualities of design, rather than a material property.

Mechanical Properties

These are measured by the material's response to stress and strain in elastic deformation. We find its tensile and compressive strengths, brittleness, and ductility by seeing how it responds to stress, how it is strained by external forces. We seek to determine how it responds to static, continuous, dynamic, or intermittent loads.

Stress and Strain

A force applied to an object causes it to change its shape by deflecting, sagging, bowing out or in. The configuration of deformation is a portrait of the stress:

$$\text{Stress} = \frac{\text{Load}}{\text{Area}}$$

Elastic Deformation

An applied load, no matter how small, always exerts stress to strain a solid object. A feather, it is said, landing on a steel I-beam, will cause the beam to deflect, and this resulting strain (it is said, although I have never seen it done) can be measured with very sophisticated instruments. When the feather is removed the strain is said to disappear. The object then returns to its original dimension.

In this example deflection is present only as long as the load is present, and its effect is reversed by simply removing the load. This reversible strain is called *elastic deformation.*

Many forces acting on a structure cause only elastic deformation—wind causing a tall building to sway, or people causing a floor to sag as they walk across it. Elastic deformation is harmless, since the structure is quickly restored to its original configuration.

Factors of safety are introduced to make sure that a structure will not be stressed beyond its range of elastic deformation. The most common measure of a material's stiffness or ability to resist elastic deformation is its modulus of elasticity E.

Plastic Deformation

Permanent or irreversible deformation is called *plastic deformation.* This type of deformation is essential to building. When building materials are formed into the desired shape we do not want them to spring back again. It would be quite

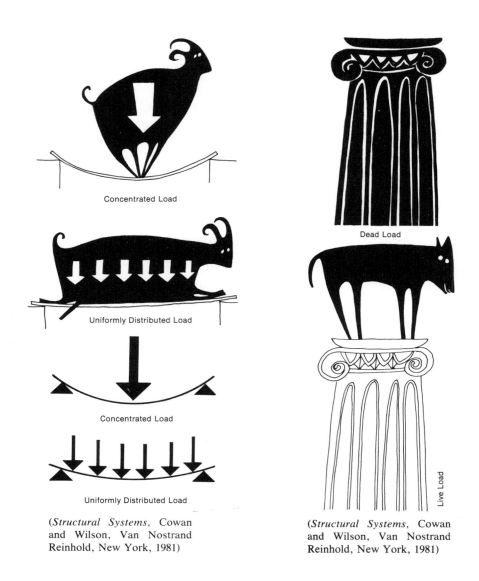

Concentrated Load

Uniformly Distributed Load

Concentrated Load

Uniformly Distributed Load

Dead Load

Live Load

(*Structural Systems,* Cowan
and Wilson, Van Nostrand
Reinhold, New York, 1981)

(*Structural Systems,* Cowan
and Wilson, Van Nostrand
Reinhold, New York, 1981)

embarrassing if steel building sections rolled into I- and H-sections suddenly reverted to their original form as a steel billet or as a rock of iron ore. When building elements are formed into structural configurations we want them to stay that way, and when they deform we want it to be an elastic deformation that returns to its original configuration. For this reason load-bearing areas of structures are made large enough so that loads applied to them will not generate stresses that cause permanent, or plastic deformations.

Stress-Strain Test

Many of the mechanical properties of a material can be accurately determined from stress-strain tensile testing. A sample, under carefully controlled conditions, is stretched to fracture, and a continuous record is made of both stress applied and resulting strain response. If the test findings are to be accepted universally by those that design and build with the material, then such testing must be conducted according to rigid standards, as they are. Test procedures for materials, ranging from structural steel to vinyl electrical insulation, are described by ASTM Standard Methods of Testing. Tables of properties for various materials are then presented in reference works, and written into building codes. They may be found in such places as the *Metals Handbook* published by the American Society for Metals.

Testing of a ductile metal will show that up to a certain point the stress-strain relationship is elastic. When the stress is removed the metal returns to its original form. Up to this point an increase in strain is accompanied by a proportionate, consistent increase in deformation. This proportionality constant—stress : strain—is indicated by *E*, the modulus of elasticity:

$$E = \frac{\text{stress}}{\text{deformation}}$$

The modulus *E* is a critical important property of a material, for it enables designers to calculate the elastic strain accompanying a particular stress or loading condition.

A comparison of the *E* values of construction materials belonging to the metals, ceramics, and molecular categories will show that metals are generally quite stiff; ceramics have slightly lower stiffnesses and the molecular materials, such as wood are the lowest.

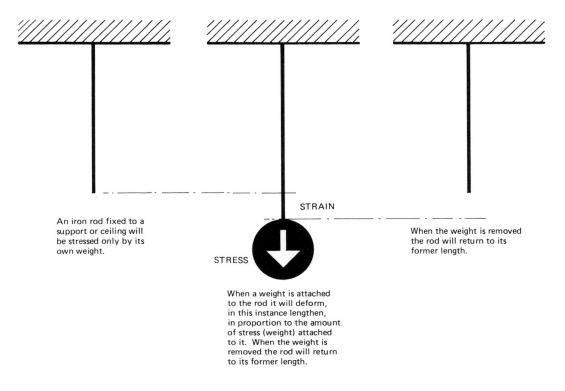

STRAIN

STRESS

An iron rod fixed to a support or ceiling will be stressed only by its own weight.

When a weight is attached to the rod it will deform, in this instance lengthen, in proportion to the amount of stress (weight) attached to it. When the weight is removed the rod will return to its former length.

When the weight is removed the rod will return to its former length.

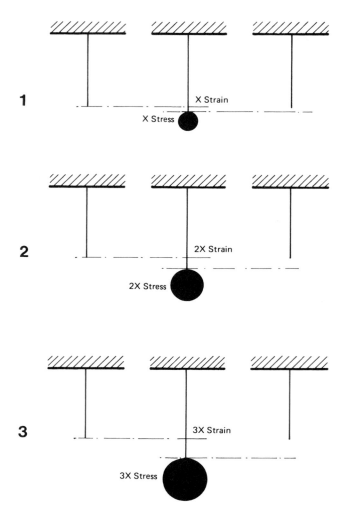

Strain is proportionate to stress when the strain is below the material's elastic limit. When the stress is removed the material will return to its original shape.

When the stress rises above the elastic limit into the area where permanent deformation appears in the body, then another condition exists entirely. If the stress is removed some of its elasticity will disappear but some plastic deformation of the material will have taken place. The limit of stress where strain forces the material behavior to move from elastic to plastic state is called the *yield point,* or *yield strength.*

Most materials do not have clearly defined stress-strain curves with exact yield points. A theoretical yield strength is calculated from the stress-strain curve, and since the yield strength marks the limit of usable strength for a ductile material this figure must be and is carefully determined. The precise definition and detailed specifications for yield point and determinations are given in ASTM standards.

When stress greater than the yield point is applied the material behaves plas-

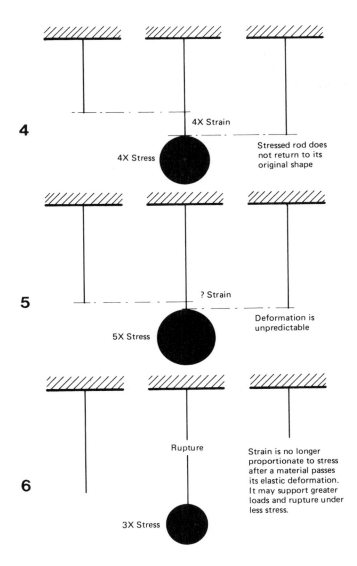

4

4X Strain

4X Stress

Stressed rod does not return to its original shape

5

? Strain

5X Stress

Deformation is unpredictable

6

Rupture

3X Stress

Strain is no longer proportionate to stress after a material passes its elastic deformation. It may support greater loads and rupture under less stress.

tically. At this level of stress the material will reach its ultimate tensile strength, elongate a bit further, and then rupture. This occurs even though the maximum load has been reduced.

If the stress-strain behaviors of four typical construction materials, structural steel, annealed copper, Portland cement concrete, and polyethylene plastic, are compared it is evident that steel has a much higher E than the other metal, copper, and will deform less in the elastic range. There is an enormous disparity between the stress-strain behavior of steel, cement, and plastics.

That the yield strength of steel is considerably higher than that of copper indicates that steel can withstand greater stresses without undergoing permanent deformation. On the other hand, steel's greater strength is paid for in higher brittleness, which means less ductility.

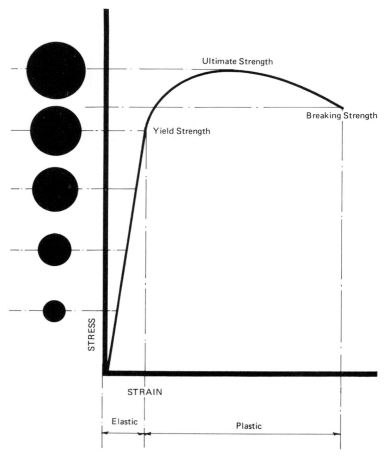

Stress-Strain behavior of a ductile Material

Metals resist permanent deformation more effectively than plastics, and their range of yield strengths can be varied by annealing, heat treating, or work hardening. It can also be influenced by the use of alloying elements. Ceramics do not deform plastically before fracture, and therefore are not generally characterized by a specific yield strength. Ceramic materials have moderately high E values and respond rigidly to elastic stresses, that is, they respond with small deformations for large stresses. They are brittle, a brittleness which is accentuated when a ceramic material is stressed in tension. This fact has limited the traditional use of concrete, brick, tile, and stone to compressive loads. The absence of an elastic-plastic transition means that yield strength, ultimate strength, and breaking strength are almost identical for ceramic materials. Materials in the molecular category are characterized by small elastic stresses producing large deflections. This may prove a disadvantage when molecular materials are used as load-bearing elements. Those molecular materials which exhibit the greatest tendency to deform elastically are the elastomers, used in rubber bands and sealants.

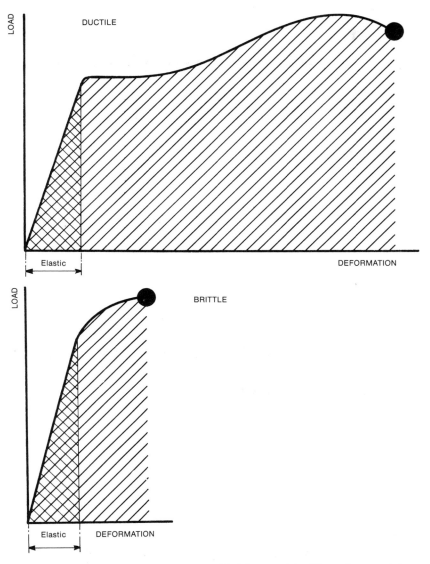

Comparison of the behavior of ductile and brittle materials. (From *Structural Systems*, Cowan and Wilson, Van Nostrand Reinhold)

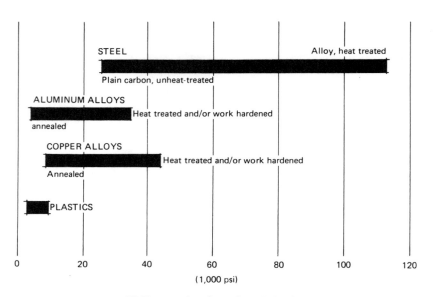

Yield strengths of metals and plastics.

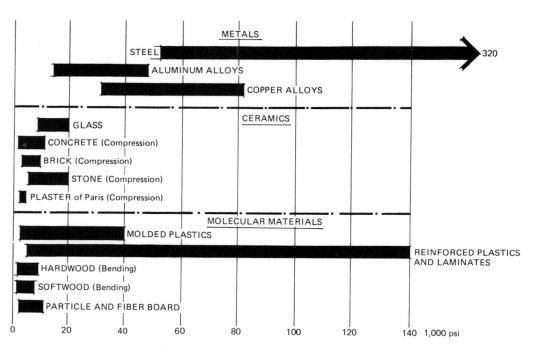

Ultimate strengths of representative materials.

Under the most severe conditions metals and plastics can be stressed to ultimate strength before catastrophic fracture occurs, whereas the breaking yield and ultimate strength of a ceramic are almost identical.

Metals are strong elastically, rigid, and fairly ductile. Ceramics are rigid, strong in compression, but brittle, with little tolerance for tension. Molecular materials are adequate for many construction applications—in fact, wood has been and remains a favorite building material in North America. But they do have limitations.

Hardness

Hardness is a measure of a material's ability to resist identation or penetration. It is determined by various tests, described in the ASTM standards in which indentors are forced into materials under carefully prescribed conditions. In general, the harder a material, the greater its wear and abrasion resistance. Materials with large modulus of elasticity and high yield strengths tend to be hard and wear-resistant. In the absence of hardness data, stress-strain data give a measure of wear resistance.

Fatigue resistance is a measure of a material's ability to withstand repeated stresses. When repeatedly stressed even at stresses below yield strength, many materials will fracture without warning. The lifetimes of pumps and other mechanical devices in a building's electrical and mechanical systems depend upon

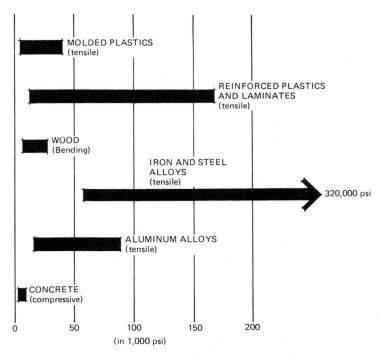

Comparative strengths of construction materials.

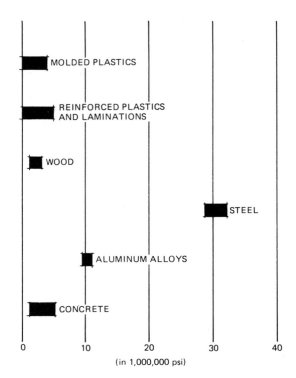

MOLDED PLASTICS

REINFORCED PLASTICS
AND LAMINATIONS

WOOD

STEEL

ALUMINUM ALLOYS

CONCRETE

0   10   20   30   40

(in 1,000,000 psi)

Comparison of stiffness of construction materials.

the fatigue behavior of the materials employed. The use of adequate shock-absorbent mountings deserves careful consideration.

Damping capacity is a measure of a material's ability to dissipate or deaden mechanical vibration. Since sound is mechanical vibration, a material's ability to absorb sound is directly related to its damping capacity.

Impact strength or *toughness* marks a material's capacity to absorb impact without fracturing. It is defined as the total energy from elastic deformation to fracture which a material can absorb before breaking under impact. Two qualities, strength and ductility, affect this property; ceramics are strong, but do not deform significantly under static or dynamic load, so they lack toughness and shatter under impact.

Plastics can stretch up to several hundred percent under load, but their strength is relatively low, hence they absorb little energy under impact. Metals, with good strength and ductility, are the toughest of the common construction materials.

Thermal Properties

When subjected to temperature changes, a material may change its state, solidify, melt or vaporize, expand or contract, and conduct or reflect heat.

Melting Temperatures

Certain mechanical properties, such as tensile strength, give an indirect measure of other properties, such as hardness. Melting temperature is a similar property. Materials with high melting points usually can be relied on to retain their mechanical properties over a greater temperature range. In addition, they tend to be stronger and more chemically inert than materials with lower melting points.

As a rule of thumb, materials with high melting temperatures, such as ceramics, perform best at high temperatures. Metals perform moderately well, and molecular materials perform least well.

Thermal Conductivity

The material's conductivity, the ability to transfer heat from a region of high temperature to a region of lower temperature, is of special importance to the designer.

Metals transport heat effectively; they typically exhibit the highest thermal conductivities. Ceramics have much lower conductivities, and molecular materials the lowest.

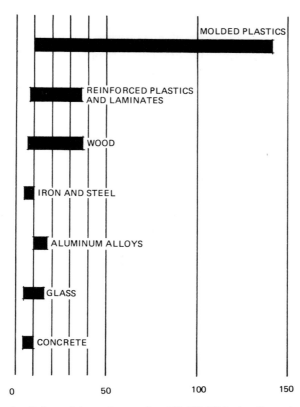

Coefficient of thermal expansion (1/1,000,000 inches for each degree F increase in temperature.

Thermal Expansion

One reason buildings seem to move is that they *are* moving, thermal changes cause materials to expand and contract. Furnace ductwork crackling as the furnace starts up is an example of thermal expansion in action.

Ceramics tend to have slightly lower coefficients of thermal expansion than metals, and plastics usually have much higher coefficients than metals. It is apparent that this ranking is inverted compared to the melting temperature ranking; thus, the lower the melting temperature, the higher the thermal expansion.

A steel bridge one mile long expands about 12 in. as its temperature is raised from 70 to 100°F.

Thermal expansion is a critical property in applications where several materials are joined. For example, aluminum has twice the thermal expansion of steel. Thus, if aluminum trim is fixed to a steel base, the uneven expansion of the two materials will cause internal stresses and elastic deformation in the assembly. Noises, as well as unattractive buckling effects, may result. The need for properly designed joints in such cases is evident.

Electrical Properties

Electrical conductivity is of interest to the building designer. It is a characteristic which is closely related to a material's thermal conductivity. Metals conduct heat and electrical energy easily. Ceramics have lower electrical conductivities and molecular materials have the lowest. This property accounts for the extensive use of molecular materials such as plastics in electrical insulation products.

Chemical Properties

The air and moisture to which building materials are exposed contain small amounts of active chemical compounds. Under certain conditions they can react with these materials, degrading their properties. Typically, ceramics resist chemical attack in practically all normal environments, and plastics resist attack from all except a few organic solvents.

The situation is more complicated in the case of metals. Metals degrade or corrode through the transport of minute amounts of electricity from certain regions, called *anodes,* to other regions, termed *cathodes.* The cathode accepts the electrons and remains intact, but the anode is degraded by the chemical reaction.

The essential elements in a corroding system are the anode, the cathode, and the current-carrying medium, the *electrolyte,* connecting them. The electrolyte is usually a solution of water and a gas, such as carbon dioxide or sulfur dioxide. Preventing corrosion, therefore, is simply a matter of removing one of the three elements from the system. For example, most paints are molecular materials and therefore poor conductors of electricity. Painting metallic surfaces not only shelters them directly from corrosive environments, but also provides a nonconductive barrier to electrical current flow.

Corrosion can occur when an entire material or given region in a material is subjected to a situation conductive to anodic behavior. Such situations develop in the presence of certain impurities in the metal acting as cathodes and when metals with different galvanic potentials are placed in close proximity. Galvanic tables ranking materials according to their tendency to become anodic can be used to predict potentially corrosive situations. Materials at the top of the list have a strong tendency to become anodic and those toward the bottom cathodic.

If a material from near the top of the list, such as carbon steel is placed in contact with a material lower on the list, such as brass, corrosion is likely to result. The steel, acting as an anode, and the brass acting as the cathode, will cause an electric current to flow. Corrosion of the steel is inevitable.

Galvanic Series of Common Metals and Alloys.

ELECTROLYTIC TENDENCY	METAL OR ALLOY
	Magnesium alloys
	Zinc
Anodic (read up)	Aluminum alloys
	Carbon steel
	Stainless steel (active)
	Lead
	Tin
	Brass
	Copper
Cathodic (read down)	Bronze
	Stainless steel (passive)
	Gold

Materials at the top of the list have a strong tendency to become anodic and those toward the bottom cathodic. If a material from near the top of the list, such as carbon steel, is placed in contact with a material lower on the list, such as brass, corrosion is likely to result. The steel, acting as an anode, and the brass acting as the cathode, will cause an electric current to flow. Corrosion of the steel is inevitable. SOURCE: ''Properties of Materials,'' in Olin, Schmidt, and Lewis, *Construction: Principles, Materials and Methods.*

Thermal Movement

MATERIAL	AVERAGE COEFFICIENT OF LINEAL THERMAL EXPANSION, MILLIONTHS (0.000001) PER °F	THERMAL EXPANSION, IN. PER 100 FT PER 100°F TEMPERATURE INCREASE (TO CLOSEST $\frac{1}{16}$ IN.)	
Clay masonry			
clay or shale brick	3.6	0.43	($\frac{7}{16}$)
fire-clay brick or tile	2.5	0.30	($\frac{5}{16}$)
clay or shale tile	3.3	0.40	($\frac{3}{8}$)
Concrete masonry			
dense aggregate	5.2	0.62	($\frac{5}{8}$)
cinder aggregate	3.1	0.37	($\frac{3}{8}$)
expanded-shale aggregate	4.3	0.52	($\frac{1}{2}$)
expanded-slag aggregate	4.6	0.55	($\frac{9}{16}$)
pumice or cinder aggregate	4.1	0.49	($\frac{1}{2}$)
Stone			
granite	4.7	0.56	($\frac{9}{16}$)
limestone	4.4	0.53	($\frac{1}{2}$)
marble	7.3	0.88	($\frac{7}{8}$)
Concrete			
gravel aggregate	6.0	0.72	($\frac{3}{4}$)
lightweight, structural	4.5	0.54	($\frac{9}{16}$)
Metal			
aluminum	12.8	1.54	($1\frac{9}{16}$)
bronze	10.1	1.21	($1\frac{3}{16}$)
stainless steel	9.6	1.15	($1\frac{1}{8}$)
structural steel	6.7	0.80	($\frac{13}{16}$)
Wood, parallel to fiber			
fir	2.1	0.25	($\frac{1}{4}$)
maple	3.6	0.43	($\frac{7}{16}$)
oak	2.7	0.32	($\frac{5}{16}$)
pine	3.6	0.43	($\frac{7}{16}$)
Wood, perpendicular to fiber			
fir	32.0	3.84	($3\frac{13}{16}$)
maple	27.0	3.24	($3\frac{1}{4}$)
oak	30.0	3.60	($3\frac{5}{8}$)
pine	19.0	2.28	($2\frac{1}{4}$)
Plaster			
gypsum aggregate	7.6	0.91	($\frac{15}{16}$)
perlite aggregate	5.2	0.62	($\frac{5}{8}$)
vermiculite aggregate	5.9	0.71	($\frac{11}{16}$)

Brick Institute of America Technical Note #18

CHAPTER REFERENCE

Olin, H., Schmidt, J., and Lewis, W., *Construction: Principles, Materials and Methods,* The Institute of Financial Education, Chicago, Illinois, and Interstate Printers and Publishers, Danville, Illinois, 4th Edition, 1980.

II. Defining Disabilities

THE ORIGINS OF DESIGN—Cracks, Movement, and Joints

The whole building is composed of parts. Joints connect the parts. All buildings move and joints are the guardian angels of movement. Cracks in building materials are due to restrained forces that have broken the chemical and physical bonds between their atoms and molecules. Materials move because of chemical, thermal, and physical changes. If the designer foresees these movements, he or she will design a joint; if they do not, nature will provide a crack.

The service life of any material during any particular usage depends on the severity of the conditions to which the material is exposed. To think of materials as durable or nondurable, strong or weak, is a fallacy. No building material is infinitely strong, all can be strained by some level of stress. To prevent failure, materials must be joined in such a way that they are not dangerously stressed.

We have some knowledge of materials and a general understanding of the mechanisms of failure. We can therefore tolerate a few minor shortcomings, but gross failure, it would seem, is inexcusable. The fact that buildings disfigure themselves shortly after completion indicates nothing as much as it does lack of respect for the nature of building materials and their environment.

Traditional buildings were the result of hand craftsmanship. They involved small components with a great number of joints. This means a small amount of movement possible in each joint. Modern industrialized building systems use large components. The number of joints is greatly reduced. The result is a comparatively large differential movement at each joint.

Traditional building also joined many materials on the site as conditions demanded. Large component industrial construction does the opposite. It is only viable if the cutting and fitting characteristic of traditional building operations is largely eliminated from the site and done under controlled conditions in the factory, which means very large preassembled pieces brought to the job site.

There is invariably a considerable difference between the intended and the actual sizes and positions of building components as they are erected. We find inaccuracies in manufacture and inconsistencies in field conditions, setting, and erection. These are, however, admittedly man-made difficulties and might somehow be avoided if we would try hard enough. But dimensional changes from thermal loads and moisture content are due to the inherent properties of materials and are unavoidable, although some claim reasonably predictable. Today's designers are concerned with joint movement, but the understanding of building and material movement is fairly limited and has only recently become an object of building research.

Photo by F. Wilson

Everyone agrees that cracks are unsightly. But they can also cause very serious problems for they can very well indicate a loss of the building's weathering quality. Cracks can open the building's external fabric to rain penetration, severe wetting, freeze-thaw breakdown, and eventually a weakening of the building structure itself.

Cracking and deformation are linked. Deformation has a number of causes: change of moisture content, temperature change, loads, and chemical action, to name some of those we know about; there are others we do not.

Many common building materials have a porous structure, absorb water readily and expand as they do so, then contract as they dry. Some materials such as concrete, mortar, and plaster, shrink initially as they harden. This initial movement is far in excess of reversible wetting and drying movement during the rest of their lifetimes, although an uneven wetting or drying after a material has set may cause considerable warpage and resulting consternation among the building's users.

Temperature change causes the material to expand as it becomes warmer and contract as it cools. Calculating the amount of deformation for a range of temperatures to which the material is exposed is fairly uncomplicated if the material is free to move and cares to conform to our calculations. However, if the material

Photo by F. Wilson

is restrained, considerable judgment and experience are required to anticipate the stresses involved, which can be of a surprising magnitude.

When stresses are greater than the strength of the material, cracking occurs. This is a simple truism. Cracking is invariably caused by restraint of one kind or another. Cracking either eliminates or redistributes stress in the material. The amount and size of cracking depends on how the new stress system adjusts to resist loads.

It is possible to avoid cracking by using a material strong enough to resist developing stress. It is also possible to control the size of cracks by reinforcing materials weak in tensile stress with materials possessing superior tensile

Photo by F. Wilson

strength. Reinforced concrete is perhaps the most common example of this strategy.

But steel mesh and rods placed in concrete do not stop cracking. Instead they restrain movement and introduce stress as a result of doing so. The concrete cracks. The stress introduced into the concrete is transferred to the reinforcement, which restrains the cracks from opening. Such cracks as do occur will probably be very narrow, but there will be a great number of them to be found by the curious in the soffit of the beam.

Stress-Strain Relationships

Since all materials respond to change of load and temperature, and some with change in moisture content, while others expand or contract in response to the chemical and physical affects of their environment, it is quite obvious that every component of a building, and the building itself, is constantly changing in size and position.

If these movements cause difficulties an investigation of them must begin with the assumption that the designer considered all movement possibilities in his and her design. The investigator then retraces what should have happened to see if it agrees with what did.

A basic premise assumed by all is that materials will obey *Hook's Law*. This states that deformations occur in response to and in proportion to loads. It follows therefore that stress and strain can be defined in terms of unit dimensions. This gives rise to the concept of a *modulus of elasticity,* which is the hypothetical stress required to produce unit strain in unit length of the material.

The known strain limits of materials, established by laboratory testing, are considered to be a valid basis for assessing the strain imposed. When a restrained material shrinks due to cooling or drying, tensile stresses and strains are caused in the same way they would be if the material were progressively loaded. If stresses increase sufficiently, tensile or compressive failures take place, and the material comes apart, cracks.

Brittle materials and those that are moderately plastic fail in brittle fractures. But, those like steel, which are highly ductile deform before they finally break. There is also a time-dependent material response which complicates failure predictions. Most highly stressed materials undergo increasing strain, even when stress is held at a constant level. This response will vary with time and level of stress. When the stress is sufficiently high another level of stress development is reached, which is the stage of *creep.*

At low stress levels creep rates are usually small, increasing for most materials as the stresses increase. Increasing strain, over time may cause significant deformations and may lead to fracture without further increase in load.

Stress and strain are closely related and may equally be used in analyzing a given building problem. If we can identify loads then stresses may be used. On the other hand if the strain imposed is known it is more convenient to compare it directly with known strain limits for the material. The phenomenon of creep is more complicated. Stress and strain cannot be related precisely unless the time history of loading is known, and this is often very difficult to establish.

If shrinkage or expansion is uniform throughout a material, changes in dimension will not cause strain. If, on the other hand, movement within the material is nonuniform, less highly strained parts restrain the stressed parts and further stresses are produced. In extreme cases cracking results.

When a solid piece of material is suddenly subjected to a cooler environment, it cools and shrinks at the surface. The more rapidly shrinkage takes place the greater is the strain caused in the material. Glass manufacturers are familiar with this phenomenon, for they anneal and cool their products very slowly. They have also found means of imposing an internal set of compressive strains in the glass by cooling its surface rapidly in such a way that the stresses are locked in, and the resulting glass is toughened.

There are significant reference levels of strain that can be used to judge the importance of kinds and amounts of movement in materials. A key figure is 0.1%

strain. This is the tolerable limit of strain in many common building materials and constructions made from them, and is a rough measure of the general dimensional stability of buildings. Strains in excess of this are usually quite serious. In many instances cracking and other difficulties will arise with even lower levels of shrinkage, because many building materials are brittle and relatively weak in tension.

Brittle materials are susceptible to stress concentrations. Ductile materials have a larger reserve strain capacity that extends beyond their normal working limits. This does not mean that all components of a structure, even though ductile, have a similar reserve, for creep and fatigue add complications. Brittleness and ductility, stress, strain, and creep are unavoidable design elements. They offer an exciting design challenge, but the stakes can be high. The result of ignoring them can range from annoyance to disaster.

REFERENCE

Baker, M. C., and Hutcheon, N. B., in *Cracks, Movements and Joints in Buildings,* National Research Council of Canada, Record of DBR Building Science Seminar Autumn 1972 NRCC 15477 (1976).

DIMENSIONAL CHANGES DUE TO TEMPERATURE

Almost every one knows that if a material is warmed it will expand, and when it cools it will contract if it is free to do so. The key is "if it is free to do so"; if not, stresses are introduced. A rigidly fixed steel bar experiences compressive stress when heated, and despite its strength may relieve the stress by buckling.

To understand the dimension of change due to temperature three things must be known:

1. The response of the material to a given change in temperature
2. The change in temperature to which it may be subjected
3. The freedom it has to change dimension in response to change in temperature.

The first of these is easily obtained. The change in length of a material free to move, following a change in temperature of one degree, is called the *coefficient of linear expansion* of the material. This is usually well established, and values are available in reference books.

There is a wide range in materials coefficient of linear expansion—7 : 1 between wood and aluminum and over 40 : 1 between wood and polyethylene. The change in length of a 10-ft length of material subjected to an arbitrarily chosen temperature change of 100°F can easily be calculated and is sometimes frightening. If we are given the coefficient figures we would find that a 10-ft-long wood stud would expand 0.024 in. Under most circumstances this is negligible. The closeness of the values for steel and concrete, about 0.08 in., is known, and this is one reason why the two can be used together in reinforced concrete. A 10-ft length of aluminum would expand about twice as much (0.168 in.), but a 10-ft polyethylene pipe would expand more than 1 in., which might be a disaster. It would also contract the same amount upon cooling 100°F. The difference between summer and winter temperatures will destroy its joints unless this movement is considered.

Vinyl siding made from polyvinyl chloride must be installed with special sliding joints or held firmly in place to resist movement. Possible movement can be nearly 0.5 in. for 100°F change in temperature.

All traditional building materials, wood, brick, stone, steel, and concrete, have fairly low temperature change deformation. Even copper and aluminum, which normally are considered highly expansive, are relatively stable compared with the new plastic materials.

Yet this dimensional flitting about is only part of the problem to be considered. The forces or stresses induced by the restraint of building material are the object of concern. In this instance it is the possible deformation related to the modulus of elasticity of the material that interests us.

To illustrate this we can select four materials with different degrees of expansion: steel, aluminum, polyvinyl chloride, and polyethylene. The modulus of elasticity for steel is 30×10^6 psi; the stress induced by fully restraining a piece of steel subjected to 100°F temperature change would be 21,000 psi. Although aluminum expands twice as much as steel, its modulus is only 10×10^6 psi, or about one-third that of steel, so it is less highly stressed at 14,000 psi. Polyvinyl chloride, with modulus of 0.2×10^6 psi would be stressed at about 1,700 psi, and when we get to the polyethylene the induced stress would be a mere 380 psi, since its modulus is down to 0.04×10^6 psi. This is only 1/75th that of steel.

If these stresses are related to the ultimate tensile stress of each material we see that the steel is at about 50% of its ultimate stress; the aluminum, although less highly stressed, is at 70%; the polyvinyl chloride at 21%; and the polyethylene at only 16%.

It may be difficult to restrain the steel fully because of the considerable forces developed, but it is not too difficult to fix vinyl siding in place despite its relatively high coefficient of expansion.

This is a very simple approach to a complex subject. The modulus of elasticity of steel may not change very much over the 100°F temperature range, although steel can suffer brittle fracture at low temperature. But the modulus of elasticity of the polyethylene will change considerably. The stress may therefore be much higher than 380 psi. On the other hand, the material will have a creep elongation, because of the slow application of the load, which will relax the stress. This may also lower the breaking stress. It is obvious that the problem may be chased round and round in circles.

The figures for coefficients of expansion, elastic moduli, induced stresses, and failure stresses are only representative values and most materials have a wide range of values. They must therefore only be considered as approximations along a line of thought to be followed when considering the effects of temperature on the dimensional changes in materials. They indicate the importance of the third item which must be considered, and that is the freedom of the material to change dimension in response to a change in temperature.

The change of temperature to which a material may have been subjected is not nearly as easy to establish as the material's probable response to temperature change.

Materials inside buildings remain at approximately the same temperature as the room air. Room air does not normally fluctuate much more than 20–30°F between summer and winter. The material is therefore maintained at a more or less constant temperature.

The effects of radiation cannot be discounted. Sunlight streaming through a window may raise the temperature of a floor slab above room temperature, but this will hardly be more than 10°F. Once a building is completed the temperature within it will remain fairly stable. But during construction, in some severe climates, the material temperature may well be 20 or 30°F below zero.

A material entirely outside a building such as a garden wall will follow the fluctuations of the outside air influenced by solar radiation. A material that forms part of an exterior building wall or a roof will naturally be subjected to a temperature range which lies somewhat in between that of the garden wall and that of the furniture inside the building.

Buildings are built among other things for the purpose of maintaining a different temperature inside from that which occurs naturally outside. There will therefore be a temperature gradient through the thickness of the enclosing building members.

Since many of the critical joints of a building are located in the walls and roof the investigation must determine, if it can, what this temperature gradient might be.

REFERENCE

Latta, J. K., in *Cracks, Movement and Joints in Building,* National Research Council of Canada, Record of DBR Building Science Seminar, Autumn 1972 NRCC 15477 (1976).

DIMENSIONAL CHANGES DUE TO MOISTURE

All building materials will expand and contract to some extent when subjected to changes in temperature many will also change dimension with changes in moisture content. In general they will expand on absorbing water and contract on drying. These movements, like thermal deformations, are generally reversible except for the initial shrinkage of such materials as concretes, mortars, and plasters. Moisture may also be an agent for other chemical and physical processes that cause unusual deformations in materials, such as the corrosion of metals or the expansive reaction products introduced into concretes and mortars.

Dimensional changes due to wetting and drying are a result of the interaction between water molecules and the material's surface in which both the concentration of water molecules and the area of surface exposure to them are involved. The greater the concentration of water molecules in the atmosphere in contact with the surface, or the higher the relative humidity of the atmosphere, the greater will be the amount of water absorbed by the surface. The larger the surface area offered by the material to the environment the more absorptive it will be.

Nonporous materials such as metals, plastics, glasses, bitumens, and some building stones have been formed by cooling or solidification from the molten state. They have no internal pores and passages and therefore offer only external surfaces for contact with water or water vapor. They absorb almost no moisture, experience no dimensional change with changing moisture conditions.

Porous building materials on the other hand are formed from particles or cells initially filled with liquids or gases. They provide large internal surfaces connected to the outside faces of the material.

A nonporous material 1 in.^{3} in volume will have a surface area of 6 in.^{2}. The same volume of gypsum plaster may have a surface area of 42 ft^2, and cement plaster 420 ft^2.

Porous materials offer large internal areas which can absorb water from the water vapor present in the atmosphere. The amount that will be absorbed for a given material will depend on the relative humidity. As the humidity increases pores and capillaries fill with water.

The relationships between relative humidity and moisture content differs between materials of different pore structure. Wood, concerete, and brick can be compared. For wood and hydrated cement, the material itself is changing because of an almost chemical interaction with the water; this is usually accompanied by a large dimensional change. Filling and emptying of capillaries does not occur at the same relative humidity during absorption and evaporation. Some capillaries will remain substantially filled down to much lower relative humidities during evaporation. Porous materials may also absorb liquid water directly through capillary action. The extent and rate of this wicking action is dependent on the degree of interconnection of pores in the material and the size of the capillaries involved.

For wood in the tangential or flat grain the change will be approximately 7%; in the radial direction (edge grain direction), 5%; and in the grain direction, only about 0.1%. Sandstone may show a change of 0.07%. Moisture content changes can cause deformations in concrete ranging from 0.01 to 0.08% for some light-weight aggregate concretes, and is about 0.04% for normal dense aggregate concrete block with extreme changes in moisture content. Bricks exhibit a change of 0.007%; marble, dense limestone, and some reinforced polyesters exhibit a change of less than 0.001%.

The full range of moisture content change from dry to saturated condition may not often be experienced by materials inside buildings. Absolutely dry conditions will seldom be encountered because the relative humidity never reaches 0% in normally heated buildings. Complete saturation of a dry or partially dry material is also unlikely to occur except in very unusual circumstances, but near saturation will be approached by cladding materials on the outside of buildings when they are wetted by rain or condensation.

Materials that are saturated when they are introduced into the building constitute a special case, such as cast-in-place concrete, mortars, and plasters. They may present problems initially because of the irreversible shrinkage during the curing stage in which some of their initial moisture is chemically combined or otherwise removed to a point which cannot be regained by simple rewetting or exposure to high humidity.

Water may be removed from a material by vapor diffusion, liquid flow, or both. The exact mechanism can be very complex, since it will depend on the nature of the pores and passages, and the chemical and physical characteristics of the surfaces.

If the transfer is by vapor diffusion, the driving force is the vapor pressure difference between the moisture in the material and the ambient atmosphere. If the transfer is by liquid flow, referred to as capillary flow or wicking, the driving force is a moisture content difference.

The removal or pick-up of water by a porous material results in a change of moisture content. If the material is thick and the rate of removal from the surface is rapid, the gradient in the material can be quite large. Dimensional changes correspond to moisture content changes. Shrinkage from drying of the surface of concrete can result in tensile stresses exceeding the strength of the material. However, the nonreversible changes in dimension of materials have one salient feature—it can be assumed that they will occur only once. In this sense they are predictable, and allowances can be made to accommodate such movements without reference to future conditions.

In the case of movements of a reversible nature, an estimate of the future environment of the material must be made and the problem of prediction is more difficult. In service, materials may be subjected to a range of conditions from wetting by liquid water down to exposure to low levels of relative humidity. Their moisture content might be predicted if the conditions of exposure were known and maintained at the known level for a sufficient length of time for equilibrium to be established. This is seldom the case inside buildings, and never the case outside.

The inside relative humidity in air-conditioned buildings may fluctuate between 30% in winter and 70% in summer. This makes it difficult to follow the often prescribed rule of selecting materials at a specified moisture content consistent with that expected in use.

Although the precise prediction of the moisture content of materials in service is difficult if not impossible, the range of moisture content which they undergo can be estimated in most situations to provide a basis for design judgment.

Those materials which are outside the thermal insulation layer in the enclosure can be expected to have moisture contents corresponding to the full range of exposures from low relative humidities to very high to actual wetting by condensation, rain, and melting snow.

For materials of the building enclosure, particularly those outside the insulation, the best assumption of dimensional changes due to moisture is that they will change from dry to wet throughout the year, and will fluctuate somewhere within these limits on a daily cycle. The extreme limits of bone dry to soaking wet may well be valid design parameters.

The established dimensional change coefficients for the range from wet to dry conditions can be used to predict the maximum movements that may be experienced. These can be modified in relation to the simultaneous movements associated with thermal expansion and contraction and other predictable movements in the development of the final design details. In many instances dimensional changes due to moisture can be a major factor in failure, and should always be considered in design, if only in a qualitative way.

If we assume that these were the conditions to which a material was subjected we may trace failure, if it occurs, back to a violation of these provisions.

REFERENCE

Handegord, G. O., in *Cracks, Movements and Joints in Buildings,* National Research Council of Canada, Record of DBR Building Science Seminar, Autumn 1972 NRCC 15477 (1976).

BUILDING DEFORMATIONS—Deflection

Buildings are not inert. They bend, twist, shorten and elongate in all directions under the influence of loads and environmental conditions of dampness and dryness, heat and cold. In some building designs it is necessary to predict as closely as possible the movements due to vertical loads, lateral loads, temperature, and creep, to ensure proper functioning and that relative movements between the frame and the cladding will not cause difficulties. These movements are complex. The following is a discussion of one of them, deflection.

Bending of structural members due to loads can result in compression strain at top and tension strain on the bottom of horizontal members. The maximum displacement of the member due to these strains is *deflection*.

Most deflections considered are those caused by loads, but similar deflection can be caused by temperature, moisture, and creep. In fact, deflection can be caused by any phenomenon that causes change in length opposite in direction or different in magnitude from one side of a member to the other. These may add to the load deflection, causing problems when neglected.

The deflection of beams, girders, joists and slabs is the most recognizable movement since it can be seen and occasionally felt.

A total list of the problems caused by excessive deflection would be quite long. The most frequent are:

- Poor appearance and function
- Cracking of walls and partitions
- Cracking of ceilings
- Vibration effects
- Structural effects.

Appearance and Function

Deflection becomes particularly objectionable when it is large enough to be seen and felt. This condition causes fear for the structural safety of the building. Excessive deflection can be a sign of weakness and should not be ignored, but

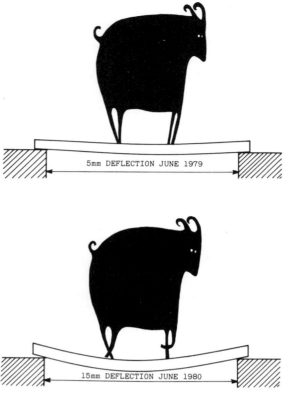

5mm DEFLECTION JUNE 1979

15mm DEFLECTION JUNE 1980

Creep. (*Structural Systems,* Cowan and Wilson, Van Nostrand Reinhold, New York, 1981)

usually the building has sufficient strength. Strength and flexibility are not synonymous.

Cracking of walls and partitions, opening of cracks at baseboard level, crushing of partition corners, and door jamming are some of the common evidences of deflection. The location and nature of cracks are dependent on the disposition of openings. A deflecting floor slab can load a wall or partition unintentionally, which may cause a flexure crack near the center, or diagonal tension cracks, or both, due to shear distortion.

Deflection of floors and roofs may also cause cracking of ceiling materials, especially plaster or plasterboard. This is most common when plaster is applied directly to the floor slab or joists. If the ceiling finish is furred or the ceiling is hung, deflection may be concealed. Segmental materials like acoustic tile are rather tolerant of deflections, for each joint helps to take up some of the movement.

In house construction, adjacent joists of unequal span have a different deflections. This may cause the ceiling to crack between them. A similar condition can occur in trussed roofs where some trusses are prevented from deflecting by partitions, while others are free to deflect.

Deflection of precast roof panels may create depressions where water will pond. This contributes to deterioration and perhaps tearing of the roofing. Cases have been reported of long flexible spans collecting sufficient water through deflection to bring about the roof's collapse.

Concrete creeps under permanent stress. Strains initially caused by loads continue to increase and deflections become larger with time. Concrete shrinks, and if steel reinforcement is placed on one side only that side will be restrained while the other shrinks freely. The result of creep and shrinkage is that the initial deflection of the member continues to grow until it eventually becomes much greater. The actual amount of increase depends on a number of factors—mix design, slump, steel size, ambient relative humidity.

Special precautions must be taken where a beam is close to and parallel to a stiffer element such as a wall. In some such instances cracks may occur under quite small deflections.

Here again the information applies principally to plaster and masonry. It appears that unless the deflection is accommodated by flexibility of finishing materials or by suitable joints, a limiting deflection of about $L/500$ is advisable to avoid problems.

More deflection problems are caused by neglecting creep and the shrinkage effects for reinforced concrete than for any other reason. Creep deflections are caused by permanent loads brought about by a member's own weight. With long spans an initial 0.5-in. creep may occur over a two- or three-year period. If a large proportion of the design live load is present the results of unanticipated creep and shrinkage deflections may be unfortunate even in short spans.

Deflections do not have to be large to cause cracking. Finishing or cladding materials attached rigidly to structural members must have enough strength to absorb loads or be flexible enough to accommodate the movement.

REFERENCE

Plewes, W. G., in *Cracks, Movements and Joints in Buildings,* National Research Council of Canada, Record of DER Building Science Seminar, Autumn 1972 NRCC 15477 (1976).

DEFORMATION DUE TO FOUNDATION MOVEMENTS

In the design of buildings it is commonly assumed that the foundation will not move. If cracks do appear in the structure it is assumed that the foundation caused them. Neither assumption is necessarily correct. The rarity of foundation failures today is due largely to the improved understanding of the properties of soil and rock materials. But detrimental movements do occasionally occur.

In practice there is much less concern about the total settlement of a building than there is about differential settlement.

Soils and bedrock are similar to any other building material in that they deform under load. But, unlike other structural materials, they must be used as they appear in nature. They cannot be controlled or influenced by a manufacturing process. Bedrock except for some unusual cases is normally an adequate foun-

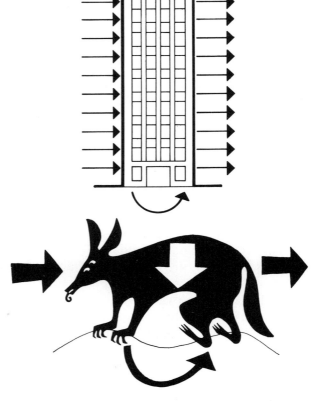

Wind pressure on the windward side of a tall building plus suction on the leeward side, produces a bending moment, and the building behaves like a giant cantilever. The problem is precisely the same as for a horizontal cantilever carrying a vertical load. (*Structural Systems*, Cowan and Wilson, Van Nostrand Reinhold, New York, 1981)

dation material. Soils on the other hand are often stressed to their limit by foundation loads.

The prediction of foundation movements is based on a knowledge of how the foundation loads are transferred to the ground and how the earth or rock materials respond to the resulting increases in stress. Common sense should tell us that there are too many variables for these predictions to be precise, but in most cases experience tells us they will be adequate.

As building loads are placed on the ground *immediate* settlement takes place due to instantaneous soil compression. Most of the immediate settlement is accommodated within the structure as it is built, and fortunately much of the differential movement occurs at this stage. Under certain conditions fine-grained soils will continue to compress under constant load for many years. This long-term compression is called *consolidation* settlement and is caused by squeezing of water from the pores in the clay.

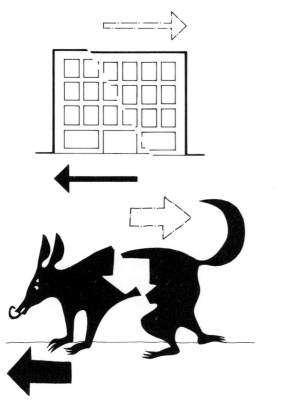

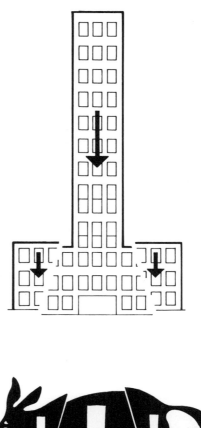

Unless the building is designed to resist the horizontal force of an earthquake it will crack and may collapse (*Structural Systems,* Cowan and Wilson, Van Nostrand Reinhold, New York, 1981)

Cracking of a building with brick walls due to the uneven settlement of the foundation. (*Structural Systems,* Cowan and Wilson, Van Nostrand Reinhold, New York City, 1981)

Differential settlement occurs for a number of reasons:

1. Local variations in soil compressibility
2. Variation in thickness of compressible soil
3. Differences in footing sizes and pressures
4. Variation in applied loads
5. Overlapping stresses
6. Differences in depth of embedment of footings.

There are three basic types of settlement: uniform settlement, tilt, and non-uniform settlement. Uniform settlement and tilt, if not extreme, do not greatly affect the structure but may cause serious problems with services and appendages such as water mains and connecting tunnels. Nonuniform settlement causes angular distortion cracks or structural failure. The amount of settlement that a building can tolerate, the "allowable" settlement, depends on its size, type, and intended use.

Large building movements can be caused by the shrinking or swelling of clay subsoil due to stresses unrelated to foundation pressures. Fine-grained clay soils may be subjected to extremely high stresses due to air drying or vegetation. Shrinkage may take place throughout the full depth of rooting and so the depth of the active layer depends both on climatic conditions and the vegetation.

Drought-resistant vegetation growing in semi-arid climates may have roots extending deeper than 20 ft. Because tree roots are extremely efficient in extracting soil moisture, both the depth and rate of shrinkage can be greatly accelerated in soils supporting such vegetation. Some soils will swell back to their original volume when they are rewetted. The up and down movement of a footing depends on the type of soil, the condition of the soil at the time of construction, and subsequent wetting and drying. Usually the worst heaving conditions develop in soils which have been previously desiccated by heavy vegetation or extreme drought and then subjected to greatly increased moisture conditions as a result of construction and irrigation practices. Greatest shrinkage usually occurs in soft, wet clay soils which have not previously been dried when deep rooted vegetation is first permitted to grow.

REFERENCE

Crawford, C. B., in *Cracks Movements and Joints in Buildings,* National Research Council of Canada, Record of DER Building Science Seminar, Autumn 1972 NRCC 15477 (1976).

ACCOMMODATION OF POTENTIAL MOVEMENTS

From all the conditions that impinge on buildings—temperature changes, changes in moisture content, loads of various forms, and movement of the foundations—it might be concluded that buildings around us are coming apart at the seams. Some of them in fact are. The seams are called joints. Sometimes there are not enough seams to come apart and the fabric of the building starts to rip off, cracks.

If there are no movements there is no need to design a joint to accommodate them. The components of the building can be fixed rigidly together, and no cracks will develop. This, however, is not practical. There will be movement; it must be accommodated.

Basically there are two methods to accomplish this. One is to resist by using strong materials. The second is not to resist: the material is left free to move and thus avoid introducing stresses. In reality both methods are used.

If we choose to resist the movement it is obvious that the material selected must have adequate strength. The potential movement to be resisted lies in the nature of the material selected, its temperature and moisture response. These must be estimated with each material.

If restraint is used so that no movement will occur this is equivalent to subjecting the material to a deformation equal to its total potential movement. This deformation can be equated to its reaction under an equivalent strain, and then, since we know the properties of the material, its stress can be calculated. If the stress is within the acceptable limits of the material then it may be possible to solve the problem in this way. Tensile stresses are quite probably accepted, but with a compressive stress there is always danger of buckling. This may lower the acceptable average stress on the material.

If the stress is beyond acceptable limits another material must be selected or the one used modified. Reinforced concrete is a good example of a modifiable material. Concrete is relatively strong in compression, but weak in tension. Embedded steel bars will take the tensile forces, creating a composite material useful for a great variety of purposes. By using steel reinforcement one or two large cracks are exchanged for a multiplicity of small ones. If even these small cracks are not acceptable the material can be changed still further by prestressing the concrete. Thus, the concrete will at all times have a residual compressive stress and will not crack.

Concrete is not the only material which can be modified in this manner. The conventional built-up bituminous roof membrane is another. The bitumen is the waterproofing component, but since it will lose some of its elastic properties at low temperature and with aging, it is desirable to reinforce it with felts of one form or another. The danger of the bitumen cracking is thus reduced.

It must be recognized that if a building component is to be prevented from moving something must hold it in place. That something will, in turn be subjected to the same forces as are induced in the component. In some instances this is not a serious problem, but in others it can lead to difficulties. A roofing membrane fully bonded to a cast-in-place concrete roof deck, for example, will not cause any problems with the deck. However, a case has been reported where a membrane applied to a deck formed by asbestos cement planks shrank with the passage of time and pulled the planks closer together, with resulting perimeter damage.

In general the method of resisting movement in a component is normally only adopted with components such as roof membranes which can be fixed over their whole area. It is seldom if ever used for components such as concrete wall panels which are mounted on the frame of the building at discrete points. Such components may have stresses induced in them as a result of differential expansions and contractions within themselves or from deflections or unanticipated settlements, but not because of deliberately introduced restraints.

Usually the building is designed to leave components free to move and not induce stresses. To be as successful as possible it is essential that components be fixed rigidly at not more than one point and that all other points of attachment be by flexible or frictionless mechanisms. The component need not be rigidly attached at any point, but might be left floating freely within set limits; a sheet of glass in a window is in such a position.

Normally the component is fixed at one location and steadied at others with allowance for potential movement, such as clapboard on a house held in place near its lower edge by nails and restrained at its upper edge by the next higher board.

Complete freedom of movement is seldom possible nor is it necessary. What is required is that the stress in the material should be kept within acceptable limits. Stress is proportional to the strain in most building materials and the strain in turn is the total deformation divided by the length over which the deformation takes place. In some situations the deformation is imposed on a material by factors which are external to it. In this case, if an adequate length of the material can be provided to accommodate the given deformation, the strain can be kept low and hence the stress will be low. Such a situation occurs where a roof membrane is called upon to bridge a joint or a crack in a roof deck. If the membrane is fully bonded to the deck the total deformation at the joint must be accommodated in the width of the joint which, in the case of a crack, can start at zero. The stresses are thus impossibly high and failure will result. On the other hand, if the membrane is only spot adhered to the deck or if a strip on each side of the joint is left unbonded, then the deformation acts on a much greater length of material and stresses are reduced proportionately.

Sooner or later one piece of material will end and a new one must be used. At such a discontinuity where relative movement takes place there must be some sort of joint. Sometimes the location of a joint is set by a change in the type of material being used. We have no option except to use a joint between the edge of the glass and the sash. When a choice can be made as to the position of a joint we should consider carefully what we are trying to achieve before selecting either the position or the type of joint.

There are two reasons for introducing a joint other than where demanded by a change in material:

1. To limit size
2. To prevent uncontrolled cracks.

Some building components can be produced in more or less unlimited length: sheet steel, rolled sections, etc. For the purposes of transportation and erection they must be cut into lengths of not greater than 60 ft. Once in place the pieces can be joined together again by welding, as is sometimes done with a girder for a bridge or a rail on a railway track.

Although batched and mixed in a concrete plant the final stages of manufacture of cast-in-place concrete takes place on the building site. With adequate resources and suitable organization concrete structures could conceivably be made in one continuous piece, such as slip formed chimneys.

Usually no more concrete is placed at one time than can be placed conveniently in one working shift. Subsequently further concrete is cast against that placed previously with no provision made for movement between the two. The reinforcement is usually continuous across the pour break and key-ways may be formed in the concrete in an attempt to ensure that no movement does take place.

These two joints, the welded one in the steel and the pour break in the concrete, are used for the same purpose: to limit the size of the component which must be dealt with at the time of construction. Such joints can be called *construction joints*. As they are not intended to accommodate movement care must be taken to ensure that they are strong enough to withstand the stresses induced by potential movement. This is usually no problem with a welded joint, but drying shrinkage in concrete may open a crack, despite the continuity of the reinforcement, because of the very low tensile bond between the new and old concrete.

Structural continuity of this sort cannot continue indefinitely and at some point provision must be made for movement. The slip-formed concrete chimney has a free end at the top and it is of little concern to anyone that the chimney grows longer or shorter. The top of a tall building is a similar situation, provided that the entire frame contracts and expands uniformly. Unfortunately this is not always the case. Parts of the frame remain in a relatively stable inside thermal condition while other parts are on the outside subjected to widely fluctuating temperatures. Care must be taken in the design of internal partitions to avoid serious cracking.

The thermal conditions to which a long but relatively low building is subjected are very similar to those for a tall one. Foundations are buried in the ground and one hopes there is no movement, and that if there is it will be small. The superstructure, on the other hand, can be less stable and could expand and contract to a considerable extent. It does not have to, if the frame is completely within the relatively stable inside conditions. Note must be taken of changes between the conditions at the time of erection and those that will prevail for most of the building's life. It should also be noted that stable inside conditions are kept stable by means of some mechanical system. Should this be inoperable for some reason then the basis for assuming stability is removed. Consideration should be given to the possibility of shutdown of the mechanical system and its effect on the building frame.

If it is decided that the extent of movement for the whole building would be such as to cause cracking of the structural frame, then a joint must be introduced at some point in the structure. This is to prevent uncontrolled cracking. It could be thought of as a deliberately introduced crack, through structure, floors, walls and roofs. Although introduced to allow for movement in either direction such joints can conveniently be called *expansion joints*.

Frequently these joints are not carried down through the basement or foundations if these are considered to be in a relatively stable environment. Where differential settlement or other foundation movements are anticipated, separate smaller buildings should be built on separate foundations.

In the absence of foundation movement it is probable that most buildings of simple layout which are not more than about 150 ft long and whose structural frames are entirely within the building enclosure do not need such joints through their frames. There is however considerable disagreement among structural engineers concerning the necessity for and the spacing of such structural expansion joints. With buildings of other than a simple rectangular plan, such as L- or H-shaped layouts, expansion joints must be introduced because of relative movement in different directions.

The so-called *control joint* is a weakening of the construction at a specific point in the hope that if movement should take place then the crack will form at this preselected location. Such a joint is less satisfactory than a properly detailed expansion joint and accommodates movement in one direction only. On the other hand it is less expensive and can be useful in concrete slabs on grade or in walls. It is usually formed by leaving or cutting grooves on one or both sides of the slab or wall and stopping some or all of the reinforcement on each side of that line.

With the exception of the structure most components of a building are made in a finite size which can readily be handled and no attempt is made to join them rigidly together. Precast concrete cladding panels are examples of such components and in most cases each panel is mounted on the structure so as to be independent of those around it. The various panels interact only when it becomes necessary to fill the gap between them for some reason, such as to keeping out the weather.

REFERENCE

Latta, J.K., in *Cracks, Movements and Joints in Buildings,* National Research Council of Canada, Record of DER Building Science Seminar, Autumn 1972 NRCC 15477 (1976).

III. Moisture Penetration and Damage

WATER, WATER EVERYWHERE

Visible water is simple to understand and cope with. Roofs are built to shed it, earth is graded to divert it, pipes conduct it. Generally water control is fairly well considered in building design.

But moisture is everywhere present as an invisible gas and this is not quite as easily controlled. There is water vapor in the air all around us. It seeps into the cavities of construction assemblies, in stud and joist cavities, into the cores of masonry units and into the fibers of wood, where it expands and rots the material.

Absorbed water is present in most construction assemblies and determines their dimensional characteristics. Moisture may also appear as solid ice within the range of temperatures encountered in most parts of the United States.

Vapor moisture is not damaging; however, it is a fickle gas, changeable, and becomes dangerous as it condenses, liquefies, and freezes. Condensation hidden over a period of time in building assemblies can cause decay in organic materials, corrosion of metals, blistering of paint coatings, and if it solidifies by freezing it can cause cracking and spalling of concrete and masonry.

Condensation and freezing may take place either as the result of a general temperature drop or the migration of water vapor to areas of lower temperature, where it condenses and freezes. Moisture problems are most likely to occur in colder climates, in heated, occupied buildings, which usually generate substantial amounts of water vapor. A normal family following its usual activities of eating, cooking, bathing, and breathing will generate as much as 25 pounds of water vapor in one day.

Condensation problems are most frequently encountered in buildings of tight construction, insulated, with occupancies and heating systems which produce humidification. The relative humidity of the air within a building is increased by moisture-producing activities. This gain in moisture content of the air increases the vapor pressure substantially above that existing in the outdoor atmosphere, and tends to drive vapor outward from the building.

Dehumidification is an important aspect of summer comfort conditioning and deliberate humidification is an integral part of typical winter heating systems. These factors create conditions conducive to vapor migration and moisture problems, unless provisions to cope with them are included in the building construction.

When water turns to vapor, it mixes with the air and occupies all the available space. In many ways water vapor acts independently of the air because its general

Fountain, Expo Montreal. (Photo courtesy of Expo Corp)

properties do not depend on the presence of air. However, when the air is either suddenly heated or cooled, the water vapor is similarly affected.

Air is saturated when the space occupied by the mixture holds the maximum possible weight of water vapor at a given temperature. The amount of water vapor necessary to saturate the air at constant pressure depends upon the temperature. The higher the temperature, the more water vapor required for saturation. If saturated air at a temperature of 40° for instance, is warmed to a temperature 60° the mixture is no longer saturated but will absorb additional water vapor. Conversely, if unsaturated air is cooled at constant pressure, a temperature will be reached at which the air is saturated. This temperature is called the dew point; if the mixture is cooled below the dew point, water will condense from the air.

The water vapor in air is called *humidity,* and relative humidity is the ratio of the amount of water vapor which a mixture contains to the amount required for saturation at a given temperature. Obviously, for a fixed amount of water vapor, the relative humidity will vary with the temperature, increasing as the temperature is lowered and decreasing as the temperature rises.

Water vapor permeability is a property of a substance which permits passage of water vapor. This property is known as *permeance*. Permeance is measured in perms. A perm is equal to one grain per square foot per hour per inch of mercury vapor pressure difference between parallel surfaces. A perm-inch is the permeance of one-inch thickness of a homogeneous material. The reciprocal of permeance is called *vapor resistance*. Values for vapor resistance of various building materials are found in charts.

As previously stated, moisture is constantly being added to the air within buildings occupied by humans or by means of humidifying systems. For this reason, the vapor pressure is usually higher inside than outside, particularly during cold weather. During hot humid days it is possible that the vapor flow might be reversed, especially when the interior is air conditioned. High indoor vapor pressures cause vapor to penetrate the interior finish and pass through the remainder of the wall to the outside. The manner in which the vapor pressure decreases from the high value inside to the low value outside determines the relative humidity at all points within the wall. The drop in vapor pressure through the wall is in proportion to the vapor resistance of the constituent parts. Vapor transfer is analogous to heat transfer.

The movement of water is largely independent of air movement. Water vapor establishes a pressure proportional to the amount of water vapor present within the air mix. Air with more vapor has a higher vapor pressure. Vapor moves through air by diffusion, without relying on air ciruclation to carry it. The critical times for condensation within or on building walls are when temperature and vapor pressure differentials across the wall are greatest.

Like other gasses, water vapor moves from an area of high pressure to an area of low pressure until equilibrium is established. During cold weather, the difference in pressure between inside and outside causes vapor to move out through every available crack and directly through permeable materials.

Moisture can cause or contribute to the breakdown of materials by chemical changes such as steel corrosion, physical changes such as the spalling of masonry by frost action, or biological processes such as the warpage and decay of wood. High humidity favors wood decay. Many insulating materials show permanent change in the course of time when in contact with water. The insulating value of most materials is greately decreased by the presence of free water in them. Volumetric changes in concrete and concrete masonry units due to wetting are relatively great and can be very destructive.

Condensation is also insidious, for it may take place over a period of time in concealed locations within walls, floors, or ceilings or roof assemblies and not be recognized until a conspicuous failure announces its presence. Air saturation is most likely to take place at low temperatures, when the risk of freezing is also a danger.

When water vapor is permitted to enter a wall assembly and condensation occurs within its outer elements, frost or ice may develop depending on the outside temperature. In weather that is continuously cold for a long period of time the frost may build back into the wall cavity or fibrous insulation, forming a mass of ice. If the outside temperature rises frequently the frost or ice will melt and be absorbed by hygroscopic materials such as wood, or it may run down nonabsorbing or already saturated surfaces to freeze again in these locations.

In masonry walls, water seepages to the outside may be harmless as long as the weather remains above freezing, but the safest course is to prevent water seepage into the building. The inclusion of weep holes and base flashings encourages runoff to the outside and bars seepage to the interior.

Accumulation of moisture within insulation lowers its efficiency, causing greater heat loss. Trapped moisture in siding and sheathing is one of the principle causes of paint failure. In seeking to escape from behind vapor-resistant paint films, moisture can cause paint blistering and peeling, especially when warmed by solar radiation. In extreme cases trapped moisture in wall cavities causes decay and rot.

The principles of vapor movement and control within roof and ceiling construction are much the same in walls. Water vapor from interior spaces moves toward the colder outside and may condense in concealed spaces of the roof and ceiling assembly.

Roofing materials such as wood shingles can "breethe" and are affected very little by trapped vapor; other materials such as built-up roofs are likely to be ruptured by excessive moisture. When a built-up roof blisters and cracks, permitting water to enter the roofing, continuous and rapid deterioration is inevitable.

Condensation within unventilated roof and ceiling spaces may collect in the form of ice, frost, or water on the rafters, roof boards, and on projecting nails. Water also may be retained between the roof sheathing and impermeable exterior roof covering such as built-up roofing. On sunny days even at low temperatures, frost or ice may melt and seep to the ceiling, below causing staining and possible damage to gypsum board and plaster ceilings. Rapid deterioration and decay of roof members also is possible in extreme cases.

Adequate vapor barriers and the ventilation of roof, ceiling spaces above the insulation are essential. If they are not provided signs of rot and deterioration will be found.

In crawl spaces, water vapor movement from the ground below is generally a greater problem than vapor movement from inhabited spaces above. Moisture in some soils can rise as much as 11 feet above the water table. This depends on ground water table, character of the soil, and drainage around the foundation. As a result of this action large amounts of water may be introduced into the crawl space by evaporation.

Excessive vapor buildup can cause condensation on the outer ends of the floor joists and other structural members which are cooled by heat loss to the outside. Water vapor from the crawl space may also enter walls, move upward within stud spaces, and even reach roof construction where the wall construction permits.

Some of the means of eliminating temperature condensation are listed below:

1. Reducing the humidity of the air. This may be accomplished by adequate ventilation if the high humidity is caused by conditions inside the building.

2. Increasing the temperature of the surface upon which the condensation occurs. Probably the simplest means of increasing surface temperature is to increase the movement of air over the surface.
3. Increasing the heat resistance of the wall. This is usually done by the addition of an air space or insulation behind the interior finish.

Walls should have vapor barriers on their warm sides and adequate flashing to divert the water that penetrates them to the outside.

REFERENCES

Olin, Harold; Schmidt, John; Lewis, Walter, *Construction: Principles, Materials and Methods,* Moisture Control, Section 104.
Brick institute of America, *Technical Notes: On Brick Construction,* McLean, Virginia (Various dates).

Moisture Problems in Buildings—Checklist

Evidence of moisture-related problems often is clearly visible to the unaided eye. The indications below will help in locating existing or potential problem areas in structures.

- Spalling or other types of frost damage
- Stains from biological growth
- Efflorescence
- Damage to floors, plaster, dry-wall, paint, etc.

Other observable effects are:

- Increased chemical reactivity of masonry surface to atmospheric pollutants
- Increased interior relative humidity

Most Likely Sources of Moisture

Look for:

- Faulty mortar joints
- Damaged parapets
- Cracks in the masonry
- Defective caulking, sealants, and expansion joints
- Defective gutters, downspouts, and flashing
- Copings with leak
- Rising damp
- Ivy growth
- Damaged surfaces, such as sandblasted brick

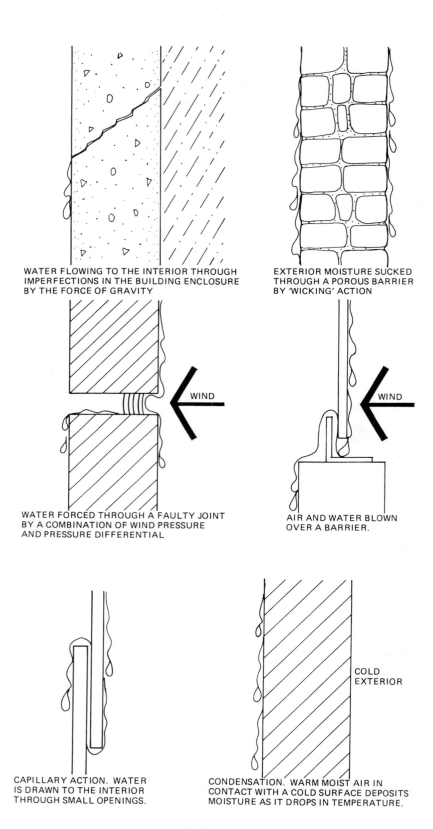

WATER FLOWING TO THE INTERIOR THROUGH IMPERFECTIONS IN THE BUILDING ENCLOSURE BY THE FORCE OF GRAVITY

EXTERIOR MOISTURE SUCKED THROUGH A POROUS BARRIER BY 'WICKING' ACTION

WIND

WATER FORCED THROUGH A FAULTY JOINT BY A COMBINATION OF WIND PRESSURE AND PRESSURE DIFFERENTIAL

WIND

AIR AND WATER BLOWN OVER A BARRIER.

CAPILLARY ACTION. WATER IS DRAWN TO THE INTERIOR THROUGH SMALL OPENINGS.

COLD EXTERIOR

CONDENSATION. WARM MOIST AIR IN CONTACT WITH A COLD SURFACE DEPOSITS MOISTURE AS IT DROPS IN TEMPERATURE.

Evaluate the various procedures available. A simple maintenance measure may suffice in some cases; in other cases, more extensive treatment may be called for. Possible remedial methods that should be considered are:

* Maintenance of roofs, flashing, gutters, downspouts, windows, caulking, mortar joints, etc.
* Damp-proof courses
* Below-grade waterproofing agents
* Porous tubes to encourage evaporation
* Weep holes
* Drainage of water table
* Water-repellent coatings

SOURCE: Bryant, Terry, Regional Operations Manager, SERMAC Division Service Master Industries, Inc.; *Technology and Conservation,* Spring 1978, Vol. 3, no. 1.

WATER INFILTRATION AND PRESSURE DIFFERENTIAL*

The key to water-infiltration-resistant design is an understanding of the effect of pressure differential on and within buildings. The air pressure difference between the interior and exterior of a building is the identifiable agent responsible for causing all building leaks.

There are several sources of pressure. The obvious are: water head or gravity and wind. The less obvious but equally important are the result of divergent air-conditioning pressures in buildings. It must be remembered that these pressures often combine to increase building infiltration problems.

To illustrate the action of pressure differential we can use the commonplace action of the soda straw as an example. When someone sucks on a soda straw they reduce the pressure in the mouth cavity making it less than surrounding atmospheric pressure. The difference in pressure between the outside atmosphere and that inside the mouth cavity pushes the soda up the straw into the mouth counter to the pull of gravity. Astronauts in space might suck on a straw without causing any action at all. This is why squeeze tubes are used as a means of emptying containers in space flight.

Similarly, in a building or other physical facility it is not the velocity of water or wind striking the building's surface which causes leaks. It is the pressure differential between inside and outside that brings the water into the building.

Water itself, even though fluid and somewhat viscous is stable and static and must have a vehicle to carry it through apertures such as a hole, slit, crack gap or other opening. The air is the vehicle for water. If a building contains a fault in water-infiltration-resistant design the flaw will be found and a leak occur if

---

* This section is extracted from a forthcoming book on water infiltration in buildings by Elmer Botsai, FAIA, Dean, School of Architecture, University of Hawaii at Manoa.

water has accessibility to the flaw and a vehicle to carry it through. If water and the flaw do not contact only air will pass through. It is the combination of air, the vehicle, and the accessibility of water to the fault that results in building failure.

The three basic causes of pressures that must be anticipated in water infiltration design are:

1. Gravity—the most critical in creating pressure on horizontal surfaces such as roofs or decks
2. Wind—most critical in creating both positive and negative pressures or the lowering of pressures, which result in pressure differentials acting on vertical surfaces
3. Water head—the most critical in creating subterranean pressure, which is pressure due to gravity.

Pressure Differential—How It Works

For the purpose of this discussion the building is divided into three basic elements critical to water-infiltration-resistant design: horizontal membranes, walls, and subterranean walls or floors.

Horizontal membranes include roofs, decks, balconies and other exterior architectural components located above grade. The critical source of pressure differential is gravity or water head. Wind pressure does not bear on the problem, excepting its action through flashing. The quantity of water is a major factor, for negative conditions are aggravated by pressure buildup directly proportional to the quantity of water standing on the surface. As the quantity of water is increased more pressure is generated on the horizontal membrane. For example, one inch of water standing on a horizontal surface will generate five pounds of static pressure, additional water standing on the surface generates greater pressure. The static pressure itself is not the cause of leaks. The situation requires a pressure differential and a flaw in the membrane. Less pressure in the building than outside acts as the vehicle to force the water through, the flaw causing building failure. If the air pressure inside the building under the exterior horizontal membrane is six pounds, or one pound greater than the outside pressure on the roof or deck above, no leak will occur. The inside air will escape upward through the horizontal membrane. If the outside pressure is greater the roof or deck will leak, for the water will be sucked throught the flaw into the building.

It is simply a basic law of physics in dealing with pressure differentials that all elements will be carried in the direction of least pressure. The water standing on the roof will be pushed in the direction of the space with the lower pressure. Unless steps are taken to correct the flaw or the pressure differential inside the building is changed to be greater than that outside as generated by the head of standing water, water will continue to infiltrate the building.

Walls are subject to the same principle. When wind blows against the building during storms and the entire exterior surface is sheathed in water, infiltration will not occur unless there is a flaw in the buildings covering and the pressure inside is less than that on the exterior. If the inside pressure is less, water will infiltrate the building through any existing flaw in the building envelope. The amount of water cascading down an exterior wall has little effect on water infiltration. It is only in the case of horizontal membranes and subterranean walls that the quantity of water produces pressure and the hydrostatic effect becomes critical.

In considering water-infiltration-resistant design of exterior walls it is important to remember that the force of gravity only becomes critical once the water has entered the building through a flaw. Then, because of the pull of gravity the problem becomes quite complex. The origin of the leak is difficult to find, since water may travel long distances before it is detected.

A classic example of this phenomenon occurred in an eighteen-story building in Texas. Severe water infiltration became evident on the first floor of the building in a total saturation of carpeting. Building occupants sloshed through water while walking over the carpeting. The leak was eventually traced through eighteen floors to its origin on the exterior wall at the ceiling level of the eighteenth floor just below the roof parapet. Once the leak was repaired on the eighteenth floor the first-floor carpeting remained dry.

The important factor in water infiltration of exterior walls above grade is not the quantity of water but the pressure differential. The wall could be fabricated of cloth and saturated with huge quantities of water and if there were no flaws in the cloth and no adverse pressure differential the interior would remain dry. On the other hand, a properly built wall subjected to a slight mist will develop significant leaks when exposed to significant pressure differential. In proof of this, tall buildings are more susceptible to leaks at the upper floors where the pressure differential is the greatest, although as water cascades down the building it increases in volume at the lower floors. It is clear then that the wind pressure which causes pressure differentials is the culprit, and not the quantity of water. On a typical wall, as wind pressure increases the potential of pressure differentials increases and in turn the infiltration rate increases.

It is critical to understand the distinction between wind load pressure and pressure differential. The two embody divergent principles. The peculiar characteristic of wind load pressure is its capacity to push a water head vertically. In contrast, pressure differential pulls or sucks the water in.

In dealing with roofs or other horizontal membranes it is well to remember that usually the only time they are affected by wind load pressures which cause water infiltration is when the integrity of flashing is involved. There is, however, another exception to the rule which deserves discussion. It relates to shingle, shake, tile, or slate roofs, or any other unit roofing material. It will be found that the basic principle of wind pressure rather than quantity of water governs infiltration.

Decades ago good construction practice indicated that shingle roofs have a 4 in 12 roof pitch. The 4 in 12 pitch gives a 4-in. vertical rise per foot, which guarantees adequate protection from infiltration even under severe storm wind loads of 80 miles per hour that will blow a 3-in.-high vertical head of water. The 4 in 12 pitch renders a safe condition because of its additional 1-in. vertical rise factor of safety. A 3 in 12 pitch, with its 3-in. rise, would be subject to infiltration during an 80-mile-per-hour wind load because no latitude is provided for a factor of safety.

It may be concluded, accordingly, that although pinhole leaks, rips, and tears in roofing systems are aggravated by water volume, shingles are not, except at gutter and eave conditions or at low slopes where water quantity becomes a problem because of inadequate pitch.

With *subterranean walls* or *floors below grade* the major cause of infiltration is the pressure on the architectural element generated by water quantity, water head, and gravity. On many construction sites it is not uncommon to find a 10-ft water head generating 600 pounds of pressure. Under these conditions correcting water infiltration problems becomes a horrendous task.

It is not only extremely difficult, it is sometimes impossible to correct a mistake made in initial design and construction. It is therefore critical that original construction be closely supervised and fabricated properly, for this is when water-infiltration-resistant design has the greatest opportunity to be successful.

In an actual example of this condition, a computer room in the basement of a commercial building developed a small leak through a crack in a wall below grade. The small leak destroyed the humidity control required for proper operation of the computers and forced the computer technicians to work with wet feet on a wet floor, a dangerous condition when dealing with electronic equipment.

The author, Elmer Botsai, first proposed a three day shutdown to excavate the exterior of the damaged wall and repair the crack properly. This was rejected due to the loss that would be sustained if the facilities were inoperable ($100,000 per day). Accordingly, to solve the problem without a shutdown an interlocking access room on the outside was devised and a small fan used to build up pressure inside the computer room. The small increase in pressure inside the room, retained by means of gasketed access doors, was sufficient to stop the leak. The lesson, from this example, is to think in terms of pressure differentials rather than volume of water.

Theory of Pressure-Differential Plane

The key to pressure differential design is the proper location of the pressure-differential plane. This is a coplanar barrier in a vertical, horizontal, curvilinear, or diagonal surface utilized to resist water penetration. There may be several pressure-differential planes within an architectural element, and each can be objectively identified. For example, in a raw concrete wall the main pressure-differential plane is the extreme outside surface. The wall's cross section reveals a series of other differential planes between the exterior surface and the interior surface, which is the major secondary differential plane of the wall section. All planes between exterior and interior are tertiary.

Although this applies to all other porous wall material assemblies as well as concrete it does not hold true for hard-surfaced, nonporous materials such as sheet steel or glass. In these materials the only major pressure-differential plane is the extreme exterior surface. For the purpose of water-infiltration-resistant design, no other pressure differential plane exists.

Importance of Location and Continuity

The building designer must objectively decide the proper location of the primary pressure differential plane to be utilized as the principal point of resistance. This decision cannot be left to happenstance, for once water has breached this plane it cannot return. It will continue to move in the direction of least resistance. Should the cause abate and no other water be sucked into the flaw, the original water will remain in the wall. It will not filter out past the initial plane.

The second major consideration is the absolute necessity of continuity of the pressure-differential plane. In areas where the plane may be breached, such as at a head, condition 6 in. of flashing and counterflashing can be installed to provide adequate defense.

It is prudent practice to locate the pressure differential plane of a wall as close to the exterior of the building as possible, for if water does penetrate the extreme outside surface and then contacts the major pressure differential plane it is easier to remove. In contrast to building practice in North America, many European builders design window and door assemblies and wall systems that allow water to penetrate the exposed surface into the building before a method is provided to remove it. Their theory is that it is simpler to allow water into the building and then remove it than to design an absolutely impenetrable system on the building's exterior envelope.

This practice somewhat resembles in theory, although it differs from it in detail, my proposal for the location of the major pressure differential plane as near the exterior of the building as possible for ease in removing water should it penetrate. The key to any pressure-differential-plane design in any location is the knowledge that water will not return to the exterior once it has penetrated the initial barrier. It will continue to travel in the direction of the least pressure, thus rendering it most difficult, if not impossible, to deal with.

Roofs

In the design of traditional roof systems there is no choice except to locate the pressure differential membrane on the outside surface. With some new roofing systems currently being developed recommendations have been made to move the principal membrane into the roof system as an integral component. However, it is generally agreed that the membrane should never be located on the inside surface of the roofing system. To reduce exposure to leaks the membrane should be near the water source.

An interesting result of the energy crisis affecting roof membranes has been the discovery that some traditional types of built up roof membranes do not perform as well as they have in the past after the addition of insulation. As a result of studies conducted by the National Bureau of Standards (NBS) it was found that insulated membranes are subject to far higher temperatures and a far wider range of temperature fluctuations than uninsulated membranes.

Fully exposed, dark-colored insulated membranes in Europe have been found to be subjected to a range of surface temperatures ranging from -20 to $+80°C$ (-4 to $+175°F$) and to rapid reversals of temperature in the magnitude of $60°C$

(110°F). In some severe weather environments such as a Alaska, tests have indicated that it might be best to place the membrane under the insulation. This would protect the membrane from severe temperature cycles. The membrane was designed for one function—to keep the building dry. It was not designed to perform the additional function of insulating the top of the roof deck. However, installation costs for such roofing systems appear to be higher, and precautions must be taken to protect the membrane from "reflecting" a crack in the deck immediately below it, which would cause the membrane to crack and leak. There is general agreement that the membrane should move up into the roofing system rather than adhere rigidly to the top of the deck surface.

It remains common practice to rely on the membrane approach, wherever the pressure-differential plane is located, to protect buildings from water infitration through its roofing system. The importance of a surface without defects is obvious. There is a tendency among some designers currently to specify huge, single-piece, prefabricated membranes and thus reduce the number of splices, which are potential weak spots in membrane systems and also reduce field costs.

According to the General Services Administration (GSA), the roof is one of the most trouble-prone components of a building. Recently over 300 lawsuits were documented because of roof problems in the United States. However, about 3 billion square feet of flat-type membrane roofs are installed in the United States annually, so the number of lawsuits may not indicate an alarming trend.

Walls

Because of the complexity of installation details it is not economically feasible to place the pressure-differential plane on the inside of subterranean walls. The exception, of course, is corrective work where no other choice exists. Exterior walls above grade offer two basic choices. Membrane should always be placed immediately under the outside weather surface or on the inside surface of the wall. For example, in light wood frame or metal stud construction, the membrane is placed under the outside weather surface but on top of the studs, whether the outside surface be plaster, shingles, wood, aluminum siding, or plywood. To move it further inside would expose the studs. When the membrane is placed directly under the outside weather surface the shingles or plaster shed about 80–90% of the water and the membrane repels the remainder. This is a reasonable decision concerning the location of the pressure-differential plane acting as the principal membrane.

When using masonry wall systems, such as cavity wall brick construction, we find traditional methods function well, with the pressure-differential plane located on the inside of the wall. It was formerly the practice with brick masonry to fur and plaster the inside with a hard-surface cement plaster or Keene's cement. This plaster skim coat acted as a pressure differential plane rather than a membrane: it constituted the application of an air-tight surface which prevented air as a medium to carry water inside.

The current fad of exposing old brick by sandblasting interior walls invites leaks because the pressure differential plane formed by the skim coat of finely applied plaster has been eliminated. A good coat of paint on the inside wall surface tends to serve the same purpose as the plaster skim coat.

If the pressure differential plane is placed on the outside there is no way that failures can be eliminated. If there are any pin holes on the exterior surface water will penetrate and move directly inside. But if the pressure differential plane, such as laminated layers of styrofoam set in a basketweave pattern using precautions at the points, is set on the inside surface no infiltration will occur. Obviously gypsum plaster on the inside surface of a concrete block wall cannot withstand moisture and water penetration passing through the porous block. Hard cement plaster or Keene's cement finish should be used.

In the case of metal wall systems the pressure-differential plane can be located as the designer chooses. In corrective work, it is easy to drill holes into the outside surface of the metal system to remove the water. This method has proven extraordinarily effective in correcting water infiltration by equalizing the pressure inside and out encouraging the water to escape to the outside. This is similar in principle to the European approach described earlier.

The major problem with all wall systems aluminum or plywood is found at the intersection of different materials and intersection of different systems and intersections of similar materials. It must be stated emphatically that in water-infiltration-resistant design the building designer must be constantly aware of the critical nature of fallibility at these points: Beware intersection, intersections, intersection!

NOTES ON THE DESIGN OF WEATHERTIGHT JOINTS

Joining two things together implies that they share building functions or responsibilities which must be continuous across the joint. Yet the joint functions as a separator of different conditions. If there is no desire to separate conditions there is no need for a joint. If a visual separation is all that is required then some form of shiplap or recess will solve the problem. If, on the other hand, it is necessary to stop the flow of air or water then the joint space must be blocked at some point. But it is difficult to achieve and maintain complete seal of a joint. The ability of the joint seal to maintain a separation between inside and outside conditions cannot be entirely relied upon. The fact that imperfections will inevitably occur must be considered in the design. This constitutes a combination of factors that tests the skill of the designer before construction and the diagnostician when a building leaks.

The joint must continue the functions of the components on either side. If they are part of the wall then the joint must continue the wall and the joint design must follow wall design. A prime requirement of wall design is that an air barrier be provided both to stop outside air from entering the building and to prevent building air from escaping past the insulation to cooler areas where condensation may take place. To control or avoid rain penetration it is desirable to shield the air barrier. The joint, just as much as the wall, must follow these principles if it is to prove successful.

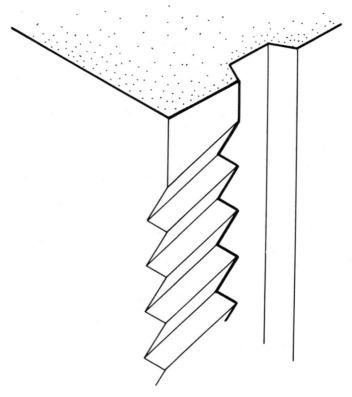

Vertical grooves in the sides of the joint will limit the movement of water towards the back of the joint apparently by providing a sharp edge which curtails this sideways flow rather than by providing vertical drainage channels as is often thought. If the sides of the joint are given a washboard profile which slopes towards the front, water that has penetrated some distance into the joint will be led back to the outside. (After J. K. Latta)

For water to penetrate a joint three requirements must be met simultaneously; (1) the presence of water, (2) a hole through which it can pass, and (3) a force directed into the building enclosure. Efforts to eliminate rain penetration must be directed either to keeping water away from any hole where there is a force which could move water inward, or to eliminate the force at any hole to which the water has access. We are unlikely to be completely successful in controlling the movement of water, and also some mistakes in design and construction are inevitable. We must, therefore, anticipate that some water will penetrate the outer skin of the enclosure. The joint system must be designed to collect this water and drain it back to the outside from whence it came at suitable points in the construction.

The design of raintight joints and an evaluation of their design when diagnosing them, can be approached in three steps:

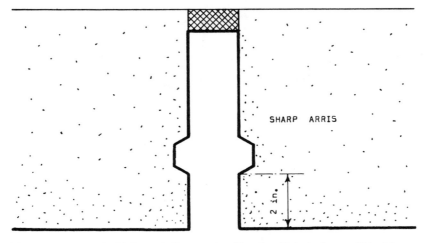

Water enters a vertical joint and clings to the sides. It will drain down within 2 in. or so, of the open face joint if the joint is sealed effectively at the back against air leakage. Vertical grooves in the sides of the joint will limit the movement of water towards the back of the joint.

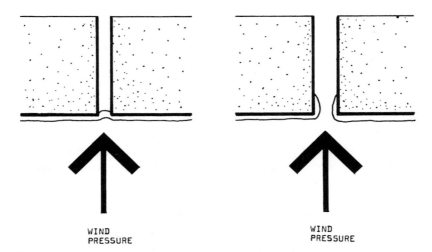

For wind pressure to be effective in pushing water inward it is necessary for the opening to be bridged by a film of water so that the wind has a plug to push against. The size of opening required to prevent this cannot be given with any degree of accuracy and will vary with particular circumstances. An absolute minimum gap width of 1/4 in. might be used in conjunction with other precautions. A 3/8 in. opening would be preferable. Larger joint widths are to be preferred. They eliminate wind pressure effects and capillary suction and do not increase the quantity of water to be dealt with to any appreciable extent. (After J. K. Latta)

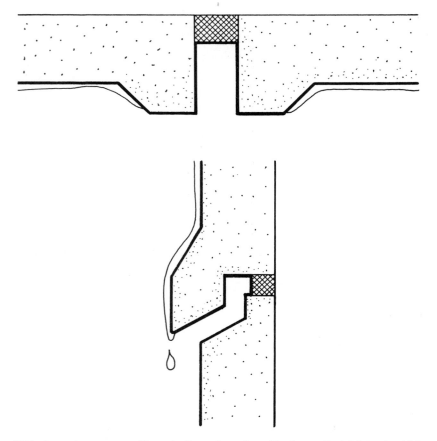

Ribbed panels or ones with projecting edges alongside the vertical joints should be effective in deflecting the water away from vertical joints. Horizontal joints can be shielded from water running off the surface above by means of projecting nosings with drips to throw water clear of the panel below. (After J. K. Latta)

1. Deflect water away from the joint to reduce the water load on it
2. Detail the joint to counteract the forces which could move the water inward
3. Drain back to the outside any water which does manage to pass the outer skin in other words, briefly: deflect, detail, drain.

Some water will land directly on the joint. But most of the water the joint must repel flows from adjacent surfaces, unless steps are taken to prevent it from doing so. As much as 70% of water on a joint surface may arrive there from adjacent surfaces.

An inside air seal is of great importance in reducing the quantity of water that can enter vertical joints. Vertical channelling to deflect water is helpful. Horizontal joints can be shielded from water by drips and projections.

The most efficient means of achieving raintight joints is to control the forces that move water inward. There are four forces: (1) momentum, (2) capillary, (3 gravity and 4) wind pressure.

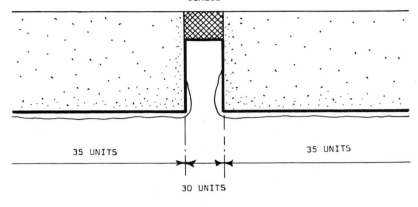

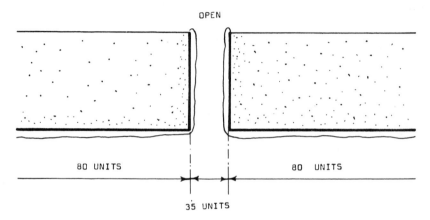

With relatively wide joint (11/16 in.) that was effectively sealed against air leakage at the inside face it has been found that 70 percent or more of the water which entered it came from the adjacent storey-height panels.

When the air seal was removed the total quantity entering the joint doubled although the directly impinging portion increased by only 20 percent. To express this difference in another way, if 100 units of rain enter the joint in the first case, 30 units will enter directly and 35 units from each of the adjacent panels. In the second cases 200 units of rain enter the joint—35 units by direct impingement and over 80 units from each of the adjacent panels. (After J. K. Latta)

Since the bulk of the water load arrives from adjacent surfaces, varying the joint width does not keep the rain out. There are limits to this statement, of course. While narrow joints may inhibit rain penetration, a joint just under 1/8 in. can cause water to rise to a height of 3/8 in. But most capillary action comes in the form of cracks of less width than this. These openings occur at defects in sealant materials, defective gaskets, between units that are in contact at intermittent positions, as well as through unintended cracks. Where the possibility of small openings of this sort can be foreseen it is desirable to limit the depth of penetration of the water by providing a larger-than-capillary space from which any water which has penetrated can be drained.

Wind pressure is an effective agent for moving water inward, but it is necessary for the opening to be bridged by a film of water so that the wind has a suitable surface to push against. Thickness of the film of water running over the face of the joint will obviously be a critical factor. If the water is shed clear of the building at frequent intervals and the sidewise movement of water is limited by vertical joints this condition can be lessened.

Water that enters a vertical joint clings to the sides of the joint and drains down within it. If the joint is sealed effectively at the back against air leakage, the bulk of the water will run down within the open faces, but the joint must be airtight. Vertical grooves in the sides of the joint will limit the movement of water toward the back of the joint.

REFERENCE

Latta, J. K., "Design of Weathertight Joints," in *Cracks, Movements and Joints in Buildings,* Division of Building Research, National Research Council, Canada (1972) NRCC 15477.

IV. Masonry

Sun, Soil, Sand—Mud Brick, Adobe

  Mud Brick

  Mortar

  Deterioration

  Signs and Sources of Deterioration

    Wind Erosion
    Vegetation, Insects, and Vermin

  Material Incompatabilities

  Patching and Repairing

  Afterthought

Then and Now—Salvaged Brick

Terra Cotta

  Common Problems

  Methods of Inspection

Clay Masonry—General Considerations

Moisture and Masonry

  Porosity

  Ceramic Glazed Brick Facing for Exterior Walls—A Special Moisture Problem

  Manifestations of Water Penetration

  A Caution—Sealers

  Flashing

Efflorescence

  Sources of Salts

  Analysis Procedure

Cleaning Brickwork

Repointing Mortar Joints

  Identifying Original Mortar in Color, Texture, and Other Properties

SUN, SOIL, SAND—Mud Brick, Adobe

The best place to begin an examination of masonry is in the study of one of the earliest and still used construction materials, mud brick. The difficulties, pitfalls, and virtues of masonry construction are reflected and made transparently clear in mud-brick construction since failings are so obvious. We also have the benefit of observing buildings and building methods that have been with us for a long time. Some say that mud bricks were first used at least three milleniums before the birth of Christ.

Failings have not outweighed successes, for the oldest surviving buildings built by Europeans in North America were built of adobe and the techniques of mud-brick construction continue to be used to this day.

Soils are the basic ingredient of mud bricks. They vary in suitability. The soil is graded for particle size ranging from gravel through sand and silt to clay. Gravel is removed, as is organic matter such as humus.

Sand resists abrasion and water damage, clay provides the strength. By mixing various proportions of sand and clay traditional builders engineered mud bricks to match structural requirements.

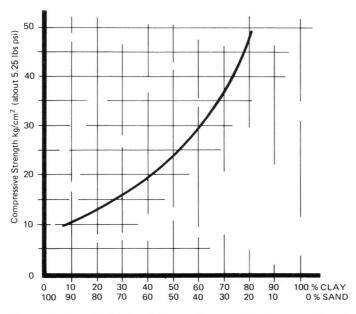

Strength of mud brick in relation to its proportional composition of clay and sand. (After Sun-dried Mudbrick-The Material AD/4/75)

Mud brick in an Egyptian Village. (Photos courtesy of Berchard M. Day)

The strength of mud bricks is low, so walls are thick. Thick walls of low thermal conductivity enclose rooms which are cooler in hot climates than those of all other materials. Thick walls deter interior temperature extremes; they tend to hold a constant temperature approximating an average of the range of exterior wall temperatures.

The greatest advantage of mud bricks has always been their availability. The cost could be measured in the amount of labor put into their production. A team of three skilled men can make from 2,000 to 3,000 bricks a day using the traditional, centuries-old mud frame methods, as their forebearers have in Egypt and other countries of the Middle East.

The difficulties are as old as the method. Mud brick must be periodically maintained; walls must be rendered, sometimes yearly. The building must be carefully protected from moisture penetration. Use of materials foreign to traditional construction practices must be carefully considered and most of the time are wisely avoided. Despite its crude appearance mud is a delicate, sensitive material that reflects and exaggerates the characteristics of other man-made masonry such as high-fired, carefully controlled bricks and steam-cured concrete blocks.

During the past few years mud bricks have been the object of considerable experiment with "improved materials." They have been "stabilized" by adding small quantities of cement, bitumen, crank case oil, and various proprietary building products. The bricks are said to be stronger and more damage resistant, but the traditional response of mudbrick to thermal expansion, contraction, and

Mud brick by Hasan Fathy, Gourna, Egypt. (Photos courtesy of Berchard M. Day)

moisture has been altered. New techniques have also altered traditional material behavior.

Traditional mud brick (adobe) construction techniques in North America have not varied widely for over three and a half centuries. Adobe building methods employed in the southwestern United States were first used in the 16th Century and are still used today. Because adobe bricks are not fired in a kiln, as are other clay bricks, they do not harden permanently, but remain in a sensitive condition of precarious stability. They shrink and swell constantly with changing water content, fluctuating with temperature and relative humidity. Generally speaking, the higher the water content, the lower the strength of the adobe brick.

Adobe does not bond permanently to metal, wood, or stone because of its unstable physical properties. Its tendency to expand and contract causes it to either crack or twist where it interfaces with other materials.

Adobe construction is sometimes combined with more stable building materials such as fired brick, wood, and lime and cement mortars. Sometimes this is the result of traditional building techniques, such as stone used for foundations and wood for roofing and lintels over windows and doors. These materials are generally held in place by their own weight and by the compressive forces of the wall or roofing material above them. The use of such foreign elements has been traditionally limited by the physical constraints of the material and they have therefore weathered successfully. However, fired brick and stronger mortars have not always fared so well in combination with adobe.

The rehabilitation of adobe buildings has proven most successful when the techniques and methods used for repair are similar to the traditional methods of construction.

Mud Brick

The adobe brick of the southwestern United States is usually molded from sand and clay mixed with water to a plastic consistency. Straw and grass are often included as a binder. Although these materials do not add any appreciable strength to the bricks they do cause the bricks to shrink more uniformly while drying. Durability of the bricks depends on the clay-to-sand ratio found in the soil. The tempered mud is placed in wooden forms, tamped, and leveled by hand. The bricks are turned out of the mold to dry as soon as they can hold their shape. They are placed on a level straw- or grass-covered surface which prevents them from sticking to the ground, and after several days are stood on end for air curing.

Mortar

Historically, most adobe walls were composed of adobe brick laid with mud mortar. Such mortar has the same physical properties as the bricks. It is relatively weak and susceptible to the same rate of moisture absorption and consequent swelling and shrinking, thermal expansion, and deterioration as the bricks themselves. Consequently no other material has proven as successful in bonding adobe bricks. Cement and lime mortars are commonly used with stabilized adobe bricks,

Mud brick by Hasan Fathy, Gourna, Egypt. (Photos courtesy of Berchard M. Day)

but cement mortars have proven incompatible with unstabilized adobe because of the difference in thermal and hygroscopic reaction of the two materials. In many instances the use of cement mortars has accelerated the deterioration of adobe walls.

Deterioration

The cardinal principle, as in all masonry repair, is simply to use the same construction techniques and replace adobe building materials with the same or nearly the same types of materials used originally. A wall in place has established its own relationship with its environment; the introduction of new materials should not unduly influence this balance. Repair and replace deteriorated adobe bricks with similar adobe bricks and rotted or termite-infested wooden lintels with a similar species and configuration of wooden lintel.

Signs and Sources of Deterioration

Damage is usually easy to detect but causes are not as readily apparent. Cracks in walls, foundations and roofs are the most common cause of concern. Some cracking, such as short hairline cracks caused by shrinkage as the material continues to dry, are normal. Extensive, large cracking indicates serious deterioration or structural problems.

Deterioration is most commonly caused by moisture. The ultimate survival of adobe buildings depends on how effectively they avoid and shed water. Erosion by rainwater and subsequent drying of adobe roofs, parapet walls, and wall surfaces causes furrows, cracks, fissures, and pitted surfaces.

When saturated by rain, adobe loses its cohesion; corners of walls and parapets slough off. If the condition is uncorrected, rainwater penetration will accelerate deterioration, eventually destroying walls and roofs, causing ultimate collapse.

Ground water below grade, due to a spring, high water table, improper drainage, seasonal water fluctuations, excessive plant watering, or changes in grade will rise through capillary action into the wall, causing it to erode, bulge, or cove.

Rainwater accumulating and standing next to walls and rain splash may erode, hollow, and cove the wall just above grade level. Coving can also be caused by spalling during freeze-thaw cycles. Water rising from the ground into the wall breaks the bond between clay particles. Dissolved minerals or salts brought up by the water can be deposited on or near the wall surface. These deposits, if they become heavily concentrated, will deteriorate the adobe.

As adobe dries, shrinkage cracks appear and loose sections of bricks and mud plaster covering crumble. Adobe walls capped with fired bricks are more water resistant. Traditional lime mortar is recommended to bond them. It is more water resistant than adobe mortar and more flexible than Portland cement mortar.

As the adobe absorbs moisture it will at a certain moisture level become soft and puttylike. As adobe becomes more plastic the weight of the roof will deform or bulge the weakened walls. When totally saturated the adobe mud will flow like a liquid. The levels of saturation that will cause this condition are determined by the proportion of sand, clay, and silt in the bricks. Serious damage is the result if wet adobe approaching the point of saturation is exposed to a freeze-thaw cycle.

Adobe walls are liable to moisture damage from shrubs, trees, or other plantings growing around the building foundation. Plant roots trap moisture and conduct it into the walls.

Wind Erosion

Wind-blown sand has often been cited as a factor in wearing down adobe walls. Evidence of wind damage is difficult to determine because it resembles water erosion. However, wind furrowing is usually more prevalent at the upper half of the wall and at building corners. Coving from rain splash and ground water usually occurs at the lower third of the wall.

Vegetation, Insects, and Vermin

Seeds deposited in the walls by wind, birds or animals will germinate as they do in any soil. Plant root action will break down adobe bricks and also cause moisture to be retained in the structure. Animals, birds, and insects often burrow or nest in walls or foundations, undermining the building's structural soundness. Termite infestation is a constant hazard, for termites travel through adobe walls as through natural soil. Wood lintels, floors, window and doors, shutters, roof framing, all are vulnerable to attack and possible destruction.

If chemical pest control is considered its potential for deteriorating adobe should be investigated. Chemicals can also be drawn into walls by capillary action and may prove injurious to humans as well as their intended victims.

Material Incompatibilities

The use of high-strength Portland cement mortar and stucco cause adobe walls, which are much weaker than these materials, to crack and crumble because of the differential expansion and contraction of the materials. Rigid steel lintels cause adobe walls to twist as they expand. Plastic and latex wall coatings seal surfaces and prevent expansion of that portion of the wall, causing portions of it to break off as a result.

Patching and Repairing

In patching and replacing adobe brick an effort should be made to find clay with a texture and color similar to the original. Fragments of the original brick, if they can be found, can be ground, mixed with water and used as patching material. Some contend, however, that spalled material may contain a high salt concentration and should be used with caution.

The same procedures used to repoint ordinary fired-brick joints can be used in patching adobe. Cracks should be raked out to a depth of 2 to 3 times the width of the mortar joint to obtain a good key or mechanical bond.

Roofs should be shored when being repaired with a new layer of mud until the new coating dries. Wet adobe is much heavier than dry and may overload the roof beams, causing them to deflect. If this occurs, when the mud dries the roof will pond in the hollow. Ponding is especially damaging to adobe roofs. Water eventually soaks through, causing wooden members to rot and the roof to collapse.

Mechanical systems in adobe buildings are a source of potential concern. Leaking water pipes and condensation are potentially more dangerous to adobe buildings than to brick, stone or frame structures because of the capillary action of the material.

Afterthought

Although susceptibility to damage, readiness to respond to change in thermal environment, susceptibility to moisture disintegration, and attractiveness to animals, plants, and insects make adobe appear to be the most delicate and transitory of building materials yet the oldest surviving buildings in the United States and perhaps the world were built of this material. Some very fine and impressive architecture has also been constructed of mud brick, from the great arch at Ctesiphon to the mission churches of Mexico and the southwestern United States to the work of Hassan Fathy. Adobe bricks are somewhat like a mafia Don: you can live with them for a long, long time if you show respect for their peculiarities.

REFERENCES

"Sun-Dried Mudbrick 1. The Material," 4/75. Nelson, Lee H., *Preservation Briefs: Preservation of Historic Adobe Buildings; #5*, Technical Preservation Services Division, Office of Archeology and Historic Preservation/Heritage Conservation and Recreation Service.

THEN AND NOW—Salvaged Bricks

The Brick Institute of America does not recommend using secondhand brick for new construction. Their reasons for not doing so constitute an informative description of brick making and use half a century or more ago. Most salvaged bricks are taken from buildings constructed 40 or 50 years ago, for it is very difficult indeed to salvage bricks laid in walls constructed with Portland cement mortar.

Although Portland cement manufacture was standardized in the U.S. after 1911, its use as a mortar did not come into widespread use until the 1920s and 1930s. Half a century or more ago manufacturing methods for brick were considerably different than those in use today. The removal of air from clay (de-airing) was unknown. Coal and wood-fired periodic and scove kilns were commonplace. The modern gas-fired tunnel kiln with accurate temperature controls was also unknown.

Generally, manufacturing conditions were such that large numbers of bricks were fired under greater kiln temperature variations than permitted today. Updraft kilns were constructed by piling the bricks themselves to form a row of arched openings in which the fires were built. Bricks were piled loosely above these arches, and, as the kilns were burnt, those nearest the fire were so intensely heated that they became partly vitrified and almost black; those at the top were only partially burned and pink in shade, with a gradual change of color between. It is from these differences of burning that the terms "arch brick," "cherry brick," and "salmon brick" originated. The extent of firing in the kiln is measured by the amount of shrinkage in the top of the pile. This type of kiln may still be used (it was during the 1950s) in small brickyards.

The burning or vitrification of bricks takes place at 1600 –2000°F, when the silicates melt and fill the spaces between the more refractory materials, binding or cementing them together. By vitrification the bricks become harder, stronger, denser, and less absorptive.

It is the nature of ceramic products to shrink during firing. Generally for a given raw clay, the greater the firing temperature, the greater the shrinkage and the darker the color.

Burning in these kilns resulted in a wide variance of finished products. Brick from the high-temperature zones were hard-burned, high-strength, durable products. Those from the low-temperature zones were under-burned, low-strength products of low durability. Temperature variations also produce a wide range of absorption and color characteristics. The under-burned bricks were more porous, slightly larger, and lighter colored than those subjected to harder burning. Those at the arch were often highly vitrified, deformed, and black.

Clays used in brick making were not as finely mixed or controlled half a century ago. When lime was present in the clay, if it was not finely divided and mixed it calcined in the burning and later slaked upon exposure to the weather. Consequently, sizable fragments occassionally expanded and chipped or spalled.

Today most bricks are produced in major brickyards. They are burned in periodic tunnel kilns under carefully controlled conditions. Burning is generally divided into six carefully controlled stages, with kilns equipped to provide a constant check on the firing processes.

Brick making half a century ago satisfied methods of construction which were quite different from those in use today. Production of both hard-burned and salmon brick was economically feasible. Most buildings had load-bearing brick walls which were at least a foot and usually more in thickness. The selected hardest, most durable bricks were used in exterior wythes. The others were used as backup. There was a great deal of sorting, grading, and selection of brick performed on the construction site by the mason. This responsibility was eventually assumed by the brick manufacturer, as it is today.

With the introduction of the skeleton frame and high-rise construction, load-bearing masonry was used less and less on major buildings. Walls became non-load bearing and gradually decreased in thickness. This evolution of building forms paralleled the production of bricks more consistent in color, size, and load-bearing quality. The demand for bricks fired in the traditional method lessoned. With the use of hollow backup units of clay or concrete masonry, the need for salmon and less rigorously burned bricks disappeared. Brick design and manufacture was concentrated on narrow, thinner walls made of bricks consistent in size and quality held together by stronger mortars that cannot easily be disassembled and reused.

The inherent danger in using salvaged brick, the Brick Institute points out, is that they will, most probably, be used for purposes for which they were never intended. Salvaged bricks usually contain many underburned units that are particularly susceptible to disintegration. As decorative interior walls or in locations where they will not be expected to withstand thermal loads, chemical attack, and the freeze thaw cycle they may be used as decoration with impunity. But to expect that they will perform as well as carefully controlled, precisely engineered contemporary masonry units in present-day curtain wall construction is asking too much of salvaged bricks.

REFERENCES

Brick Institute of America, Technical Notes #15.
Parker, Gay, and McGuire, *Materials and Methods of Architectural Construction* (3rd Edition), John Wiley & Sons, N.Y., 1958.

TERRA COTTA

Terra cotta is made from controlled, aged clay, mixed with sand or pulverized fired clay. Terra cotta is an enriched molded clay brick or block molded and fired at high temperatures to a hardness and compactness usually not obtainable with brick. The term is Italian, literally "cooked earth." Clays used in its manufacture vary widely in color, ranging from red to brown to white.

Terra cotta was most usually hollow cast in open-backed blocks with internal compartments called *webbing* which strengthened its load-bearing capacity.

Terra cotta developed its own unique technology and installation methods. It was originally used for sewer pipes, as the fireproof cladding of metal frame columns and beams, and the light floor vaulting between them. It did not emerge on the exterior of buildings until the 1880s, according to Dixon. It was glazed in creamy white and delicate pastel colors and used for almost every kind of building from coast to coast. Its use declined by the 1930s and it was succeeded by ceramic veneer in the 1940s. (Ceramic veneer is a material similar in composition but limited to forms that can be produced by extrusion.) Terra cotta tile has vertical voids like those of concrete block instead of the egg crate internal structure of the older terra cotta.

The major designs of architectural terra cotta, either cast or modeled, had to be divided into sections weighing several hundred pounds. Internal hollow portions were designed for structural integrity and to control a 6–10% shrinkage during firing. The glazes of adjoining pieces had to match and adhere firmly to the surface through firing and weathering. Units had to be meticulously dried before firing and fired without cracking or rupture.

After they had been fabricated, terra cotta units were tied to their structural supports with metal anchors. Mortar had to be carefully applied to ensure that

Terra cotta facade, New York City. (Photo by F. Wilson)

the units were uniformly loaded. They were backfilled for fireproofing and the protection of concealed construction elements. Terra cotta was thought to be fireproof and waterproof. It was easy to maintain, it did not need painting, and periodic washings restored it to its original appearance.

Although many buildings survived well a lack of foresight or understanding of

Terra cotta sculptured band around the Pension Building, Washington, D.C. (Photo by F. Wilson)

the nature and the limitations of the material, and of the nature of the building materials surrounding it, has in many instances resulted in serious deterioration. Terra cotta experiences many problems similar to those encountered by brick and stone, but also has many properties radically different than those encountered by traditional masonry materials.

This presents a serious problem. Glazed architectural terra cotta, because of its past popularity, comprises a major portion of many of our urban environments. Its infinite varieties of modeling, casting, and glazing have hidden this fact from many casual observers. Architects and owners are often surprised to find what they thought to be a granite or limestone building was in reality glazed architectural terra cotta.

Historically terra cotta has been used in two different structural systems. First it was employed as part of traditional load-bearing masonry walls in buildings of modest height. Second, it was used as a cladding material in high-rise construction. In this application terra cotta installations demanded an extensive metal anchoring system for attachment to the superstructure. The system was more complex but somewhat analogous to the precast concrete panel systems of the 1960s and 1970s.

In low-rise buildings the anchoring system was fairly simple and has remained relatively trouble free. In the high-rise building deterioration is often severe.

A diagnosis of the deterioration of architectural terra cotta is difficult because of the design and nature of materials and methods of attaching them were infinitely complex. Where terra cotta has been used as a cladding material deterioration is accelerated when the first defense, the exterior building surface, has been breached. The entire system—glazed units, mortar, metal anchors, and masonry backfill—is vulnerable. In no other masonry system is material failure potentially so complicated. Problems are compounded by the historic attitude toward architectural terra cotta as a highly waterproof system that did not need flashing, weep holes, or drips. This assumption has proven untrue. Serious water-related failure was evident early in the life of many glazed architectural terra cotta-clad buildings.

Installation errors, poor maintenance of mortar joints, and flashing difficulties eventually result in spalling and crumbling of units. Most terra cotta failures currently in evidence are in structures dating from the period between 1895 and 1920, after glazed terra cotta had become common, Dixon claims. Theodore H. M. Purden says that earlier examples seem to have survived intact. When a glaze, an impervious finish, is applied to a highly porous base, moisture that finds its way inside will tend to break down the bond between these uncongenial layers and the result will be spalling.

After the turn of the century terra cotta was used in a number of daring ways, Dixon notes. Flying buttresses at the top of the Woolworth Building required some replacement of units only two years after their erection. It was also during this period that terra cotta was applied as a continuous cladding for buildings of 15 stories or more. It was firmly anchored to masonry backup walls with no relief for either settling or thermal stress. The result has been cracking patterns characteristic of incompatibility between structure and cladding. Yet despite ignorance, many terra cotta façades remain in remarkably good condition.

There were four types of applications of terra cotta units:

1. Brownstone, a dark red or brown block which is either glazed or natural. It was hollow cast and used generally in conjunction with other masonry imitating sandstone, brick, or brownstone. It is found in Gothic and Romanesque Revival architecture in ornamental detailing such as moldings, finials, and capitals.

2. Fireproof construction, which was developed as a result of the appearance of the high-rise iron and steel frame building. It was inexpensive, lightweight, and fireproof. The blocks were rough finished, hollow, and formed to span betwen I-beam members in floor and ceiling construction. They were also used for walls and column fireproof enclosures. Certain variations of these blocks are still in production today, although not as widely used as they were previously.
3. Ceramic veneer was developed during the 1930s and is still used extensively in building construction. Terra cotta ceramic veneer is not hollow cast but is a veneer of glazed ceramic tile ribbed on the back like bathroom tile. It is frequently attached by a grid of metal ties.
4. Glazed architectural terra cotta was the most complex development of terra cotta as a masonry building material in this country. The hollow units were hand cast in molds, carved, or modeled and heavily glazed. Terra cotta casting was often used as an imitation of stone. It was sometimes called *architectural ceramics,* and was developed and refined during the first two decades of the 20th Century.

Common Problems

No two cases of terra cotta deterioration are identical. The infinite number of variations possible, including the material's manufacture, installation, number of component parts, various repairs over time, and the numerous possible causes of deterioration, make diagnosis extremely difficult. However there are some general patterns that emerge.

Failure is most commonly water related. Less frequent, although no less severe, is faulty original craftsmanship. There has also been stress-related deterioration, which is not surprising since often no provision was made for thermal movement. Damage caused by poorly considered alterations, additions or repairs is also common.

Water-related deterioration is the principal source of difficulty with terra cotta materials as it is with most building materials. The units are highly susceptible to glaze cracking, spalling, and material loss due to moisture. Sometimes entire units are found missing because of corrosion of the metal anchors behind them. Crazing, the formation of small random cracks in the glaze, is not unusual. When the new terra cotta units come from the kiln after firing they have shrunk to their smallest possible size. With the passage of time they expand as they absorb moisture from the air. This expansion may continue for many years. The glaze is in tension, since it has less capacity for expansion than the porous tile body. When the strength of the glaze is exceeded it cracks (crazes). Crazing is not unlike the random hairline cracking found on the surfaces of oil paintings or concrete surfaces that have dried too rapidly. Unless the cracks visibly extend into the porous tile body beneath the glaze, crazing is not regarded as a serious material failure. If crazing does extend into the tile it increases the water absorption capability of the terra cotta unit, which will cause it to expand further.

Spalling, the partial loss of masonry material, is caused by water trapped within

Glazed terra cotta lintel and sill, Millbrook, N.Y. (Photo by F. Wilson)

the masonry system itself. Trapped water has numerous causes; detailing in the original design, insufficient maintenance of joints, rising damp from below or a leaking roof from above. In most instances trapped water tends to migrate outward, through masonry walls, where it evaporates. In glazed architectural terra cotta, the water is held below the surface by the impervious glaze, which acts as a water barrier. The water is stopped until it builds sufficient pressure, in the presence of fluctuating temperatures, to pop off sections of the glaze; this is glaze spalling. Water pressure may also precipitate wholesale destruction of the unit itself, or material spalling.

Glaze spalling may appear as small, coin-sized blisters where the glaze has ruptured and exposed the porous tile body beneath. This may be found on several spots on the surface, or in more advanced cases in the wholesale disappearance of the glaze. Glaze spalling may also be symptomatic of the corrosion of the internal metal anchoring system which holds the terra cotta units to the larger

Crazing of terra cotta glaze at sill. (Photo by F. Wilson)

building structure. The increase in volume of the metal as it corrodes exerts internal pressures on the terra cotta unit, which in turn may spall the glaze or in more extreme cases precipitate material spalling.

Material spalling is dangerous. The visual integrity of the detailing is impaired, but more seriously some of the porous underbody, webbing, and metal anchoring may be exposed to further water entry and consequent accelerated deterioration. The units may loosen and fall as a consequence. Missing units create a passage for more water to enter the system and corrode metal anchors, which is the most serious and difficult form of deterioration to locate and diagnose. All too often the damage is severe before it is detected. Water entering the system can rust, substantially weaken, or completely disintegrate these elements. Where water has entered some deterioration has most probably taken place. Deterioration may be detected by staining and material spalling. Total deterioration and the resulting loss of anchorage will result in the loosening of units, and threatens the structural integrity of the building.

As water has always been the primary destructive force, mortar has always been the key to the survival of masonry systems. This is particularly true in the case of glazed architectural terra cotta because of the fragile nature of the system. Mortar deterioration results from all the reasons it does with other masonry systems: improper workmanship, air and waterborne pollution, expansion of the units, and the rest.

The deterioration of materials adjoining the glazed architectural terra cotta

Terra cotta cornice in Plainfield, N.J. Note crack between cast members which opens interior to water infiltration. (Photo by F. Wilson)

(flashing, capping, roofing, caulking around windows and doors) is often responsible for its deterioration. When adjoining building assemblies fail, water-related deterioration inevitably follows.

Deterioration caused by stress in high-rise buildings is not uncommon in terra cotta construction. An understanding of differential movement and stress-relieving details, such as flexible joints, occurred late in the development of American building construction. Consequently most early continuously clad high-rise buildings dating from 1900–1920 had little or no provision for normal material and building movement in the original design. The development of large stress-related cracks is often caused by unaccommodated building fram shortening under load, thermal expansion and contraction of the façade, and moisture expansion of the glazed architectural terra cotta units themselves. Cracks running through many units, stories, or large areas of material are dramatic portraits of stress-related problems.

Because it is difficult or impossible to replace damaged terra cotta units repairs have been made with brick or cementitious materials, such as stucco or fiberglass. Some are appropriate, others are not. Improper anchoring or bonding of repairs and visual incompatibility have, with the passage of time, become rehabilitation problems. Alteration damage also occurs as a result of the installation of building additions such as signs, screens, marquees or bird-proofing. Bored holes or cuts made in the glazed architectural terra cotta to anchor these additions to the building frame beneath often become sources of water related damage.

Methods of Inspection

Procedures for checking the condition of terra cotta are almost as painstaking as the production of the units. Most serious deterioration can be detected by visual inspection. This must be a very close observation to reveal hairline cracks, which are potentially as serious as advanced spalling.

Tapping. This is a somewhat inexact method of detection but nevertheless one of the most reliable presently available. Tapping is done by striking each unit with a wooden mallet. When struck, an undamaged unit gives a pronounced ring, indicating sound internal condition. Deteriorated units give off a flat, hollow sound, indicating internal damage.

Infrared Scanning. This procedure is in the experimental stage. All materials emit heat. Infrared is invisible to the human eye but can be measured by scanning. Infrared photography, a form of infrared scanning, has been of use in detecting sources of heat loss in buildings. Broken or loose internal terra cotta pieces have less firm attachment to the surrounding elements and therefore have different temperatures. These temperature differences can be detected by the infrared scan and serve as a fair indication of internal material deterioration.

Sonic Testing. Sonic testing has been used for some time to detect internal cracks in concrete members. It can also reveal internal failure in glazed architectural terra cotta. Sonic testing registers the internal configuration of materials by penetrating the material with sound waves. The patterns are read by "bounce back" from the original source of the sound. Readings at variance with those from undeteriorated material indicate collapsed webbing or pools of water in the interior.

Metal Detection. Metal detectors indicate the presence of metals by electromagnetic impulses. They are generally useful for locating the position of internal metal anchoring. Impulses are transmitted onto an oscilloscope where they may be viewed or are converted to sound patterns heard by the operator. Original drawings are useful in predicting where internal metal anchoring should be. Metal detectors can confirm their position or the contractor or architect can locate the metal anchoring without original drawings. No reading where an anchor would be expected indicates a missing or deteriorated anchor. Information produced by metal detection is, at best, a rough approximation. It is, however, the most viable way of locating the internal metal anchoring without physically removing the units.

Laboratory Analysis. Carefully selected and removed samples of terra cotta can reveal moisture problems or the penetration of damaging salts. Much of this effort is aimed at determining causes of deterioration so that damage to sound units can be avoided as well as at determining where materials must be replaced. Laboratory analysis can also find glaze absorption, permeability of glaze adhesion, and evaluate material porosity.

REFERENCE

Dixon, John Morris, "Recipes for Baked Earth," *Progressive Architecture*, (November 1977).
Tiller, de Teel Patterson, *Preservation Briefs: The Preservation of Historic Glazed Architectural Terra Cotta #7*, Technical Preservation Services Division, Heritage Conservation and Recreation Service.

CLAY MASONRY—General Considerations

The clay, the method of manufacturing, and the degree of burning affect the compressive strength of clay masonry units. With some exceptions, the plastic clays used in the stiff-mud process have higher compressive strengths when burned than clays used in the soft-mud process or dry-press processes. For a given clay and method of manufacture, higher compressive strengths are associated with higher degrees of burning. The compressive strength of brick varies from 1500 psi to more than 20,000 psi, due mainly to the wide variation in the properties of the clays used.

The absorption of water is also dependent on the clay, manufacturing method, and degree of burning. Plastic clays and higher degrees of burning generally produce units having lower absorption. Generally, the stiff-mud process produces units with lower absorption than either the soft-mud or dry-press processes. Suction, the initial rate of absorption of brick, is caused by pores or small openings in burned clay that act as capillaries and tend to draw or suck water into the unit. Suction has little bearing on the transmission of water through the brick itself, but it does have an important effect on the bond between brick and mortar. Maximum bond strength and minimum water penetration are obtained

Facade of Pension Building from subway entrance Washington, D.C. (Photo by F. Wilson)

Interior of Pension Building, Washington, D.C. The columns supporting the series of arches are made of brick and are hollow. They are purely decorative, the roof is supported by a series of metal trusses. (Photo by F. Wilson)

when the suction of the brick at the time of laying does not exceed 20 grams of water (0.7 oz.) per minute. Brick having suction of more than 20 grams per minute should be wetted prior to laying in order to reduce the suction.

The durability of clay products results from fusion during burning. The only action of weathering that has any significant effect upon burned clay products is alternate freezing and thawing in the presence of moisture. Where the annual average precipitation exceeds 20 inches and alternate freezing and thawing is common, the ability of brick or tile to resist freezing and thawing without disintegration is very important. For brick and tile produced from the same raw material and by the same method of manufacture, either high compressive strength or low absorption are fairly accurate measures of suitable resistance to freezing and thawing.

For the same raw materials and methods of manufacture, lighter colors are associated with underburning, producing salmon brick from clays which burn red. Underburned brick is softer, more absorptive, and has decreased compressive strength. Overburning produces the clinker bricks, dark red to black in red clays and dark speckled brown in the buff clays. Burning at higher temperatures tends to produce a harder brick, decreases absorption, and increases compressive strength.

Expansion due to moisture of well burned clay products can occur over long periods of time. Currently, a coefficient of moisture expansion of 0.0002 is recommended. However, alternate freezing and thawing of some clay products when saturated may lead to eventual disintegration. If soluble salts are present in the masonry, moisture caused by condensation may also contribute to efflorescence.

REFERENCE

Olin, Schmidt, Lewis, *Construction: Principles, Materials and Methods,* The Institute of Financial Education Chicago, Illinois, and Interstate Printers and Publishers, Danville Illinois, 4th Edition, 1980.

MOISTURE AND MASONRY

It is generally agreed that the durability of masonry depends primarily on its resistance to the penetration of moisture. The source may be wind-driven rains or interior exposures from occupancies creating high humidities. These could be air conditioning with humidity control, food processing, or unventilated space heaters among other causes. Differences in humidity between inside and outside air will cause vapor flow within the wall and unless controlled by vapor barriers or ventilating, may condense within the wall.

Water is unwelcome in masonry walls. If a wall is saturated with water, freezing and thawing may cause cracking, crazing, spalling, disruption, or disintegration. Excessive moisture can lower a wall's thermal insulating efficiency and attack paints, plaster and other finishes. Without water, efflorescence cannot occur.

Water, however, is in abundant supply, rain and snow cover buildings and airborne water vapor is always present. Although water cannot be eliminated from walls, good design can restrict its presence so that for all practical purposes a wall will be completely dry. If the wall leaks we can only assume that it has been improperly constructed.

Waterproofing and damp-proofing precautions are taken during initial construction. These terms are often misused. *Damp-proofing* refers to a reduction in moisture penetration by capillary action. *Waterproofing* refers to treatments intended to stop the flow of water through a wall. Both should be done.

Moisture and masonry. (Photo by F. Wilson)

When moisture passes directly through clay masonry walls, it invariably does so through mortar joints. Under normal exposures, it is virtually impossible for significant amounts of water to pass directly through brick or tile units themselves. Highly absorbent brick or tile may take in water, but certainly not enough to contribute to an outright flow of water through a wall.

In efforts to preserve masonry preservatives are sometimes considered. The ideal preservative material for building materials should have the following properties:

1. It will not change the stone's appearance.
2. It is elastic.
3. It is impermeable to liquid water, but not to water vapor.
4. It is erosion resistant but will not form a surface crust.
5. It penetrates sufficiently.

6. It is chemically nondamaging to the masonry itself.
7. It is economical.
8. It will preserve the material indefinitely.

Although technology has made strides and new chemical formulations have been developed to protect masonry from water penetration, an ideal agent that performs all of the above functions is not available and is not likely to be found in the near future. Terry Bryant asks, "Is such a product necessary?"

Are water repellent coatings the answer for the ills of masonry water infilatration? Many claim that these products can prevent efflorescence, fading and dripping of paint, deterioration of mortar, spalling of masonry, water absorption in bricks, masonry surface staining, and similar disorders. Others argue that water penetration and water related problems are principally due to poor building maintenance, and that preservatives would not be needed if leaking gutters and downspouts, deteriorated mortar, rising damp, etc. were taken care of, or better yet, not allowed to reach a state of disrepair. Both positions appear to have validity.

Porosity

Knowing the properties of masonry materials is of assistance in selecting the most appropriate masonry preservation method for a specific structure.

Water, no matter what its form—humid air, rainwater, or ground water—is the single most destructive element that masonry encounters. The water-absorption capability of masonry is extremely large. Many bricks will absorb 155% of their own weight in water. Theoretically a brick weighing 5 lb can absorb one-half pint of water. In terms of a building 100 ft wide by 45 ft long by 15 ft high, the structure's four walls could hold over 2,000 gallons of water. Aside from the increased weight of the building, the potential for damage to interior finishes becomes significant.

Abrasive blasting to exterior surfaces of brick or terra cotta may make the exterior surfaces much more porous and more prone to massive water penetration.

The mechanisms for water transport are many and are dependent upon the water state, but all can be correlated with the masonry's capillary system.

Moist air readily diffuses into dry materials. There is a known relationship between the relative humidity, of the air and the amount of water introduced into various building materials. Since there can be a significant variation in ambient air relative humidity during the year, the water content of masonry can fluctuate and damage due to water absorption/evaporation cycling can result.

When the temperature at the material's surface is lower than the dew point of the ambient air, or when the interior temperature reaches the dew point, condensation will occur in the capillary channels within the masonry units. Therefore, if there are water vapor barriers on the outer surface of a wall, condensation risks are increased when temperatures inside are higher than outside.

Liquid water, primarily in the form of rainwater, can be transported in several ways, once in contact with the masonry unit. However, the major mechanism of its penetration is by capillary suction, either from a surface film or upon penetration of cracks and crevices. The amount of water and time for passage, in this case, is related to the structure of the capillary system—channel size, interconnections, etc.

The specific construction also plays a role in water passage through a wall. For limestone walls, penetration is governed primarily by the permeability of the stone itself, while in a brick wall, the joints may be five to ten times more permeable than the bricks themselves.

Vertical transport, or rising damp, also may be caused by capillary suction via contact with ground water, as well as by ground moisture infiltration.

Ceramic Glazed Brick Facing for Exterior Walls—A Special Moisture Problem

It is generally agreed that most disintegration of masonry units results from water in the masonry and is due to crystal pressure of (1) ice forming in the pores of the masonry, or (2) migration of salts and subsequent crystallization within the pores of the masonry. It is also a well established fact, confirmed by numerous laboratory tests, as well as extensive observations of masonry buildings, that the principal cause of rain penetration through brick and tile walls is openings between mortar and masonry units. Also, it is well established that much of the water that enters a masonry wall through such openings, by absorption of the masonry or as a result of condensation within the wall, escapes by capillary action and evaporation from the face of the wall.

The process, by which moisture is eliminated from the masonry has for years been known as *breathing*. Breathing is an important factor contributing to the durability and resistance to water penetration of masonry walls.

Moisture entering the walls must be eliminated before it causes trouble. It usually enters as rainwater, through the pores in building materials, into incompletely bonded mortar joints, around copings, sills, and belt courses, as condensation of vapor from the interior of the structure through capillary contact with the ground. This action is more insidious and far-reaching than most people realize.

Without water in the wall there would be little or no problem of masonry disintegration and disruption.

The ways to rid the wall of moisture are through continuous cavities within walls (hollow wall construction) with adequate weep holes at various locations in the wall, and by evaporation through the exterior of the masonry.

When exterior walls are faced with ceramic glazed brick or when an impervious coating is applied to the exterior face of walls, water cannot enter the walls through the pores of the brick nor can it escape by this means from the face of the wall. For this reason, it becomes especially important to reduce the water entering the wall to an absolute minimum and to provide positive means of escape for water that permeates the wall, either as rain or as condensed vapor.

Manifestations of Water Penetration

Once water has been absorbed onto and into masonry, several types of deterioration can occur.

The condensation of water within the capillary structure presents one set of problems, frost damage being a prominent part of this deterioration. That is, if the temperature drops to below 32°F, ice crystals can form. This formation process can result in internal pressure which, in turn, can cause spalling and cracking.

Biological deterioration is another cause of concern, since the one prime requirement for all biological growth on masonry is the presence of moisture. For

Window, Pension Building, Washington, D.C. (Photo by F. Wilson)

instance the growth of ivy along and into masonry creates a means of retaining even more water. Plants will shade the masonry surface, slowing evaporation rates as well as causing crevices to be opened for water to enter. In addition, surface growth of fungi and algae present an objectionable appearance, help entrap more moisture, and may result in permanent staining.

A third moisture-related problem is caused by the chemical reactions between surface water and air pollutants. The compounds formed may be reactive products such as sulfuric acid and carbonic acid. Carbonate-containing materials, such as limestone and marble, are susceptible to attack by such acids, the result being pitting and erosion of the material's surface.

Still another major cause of weathering of masonry involves water and soluble salts. Frequent dry-wet cycles with varying relative humidity of the ambient air will lead to an oscillating front of liquid water within the masonry. This is accompanied by periodic crystallization and dissolution of salts. The surface-level appearance of these soluble salts, i.e., *efflorescence,* usually is obvious— the white deposits on bricks, etc. Below-surface salt formation, *subflorescence,* is not as easily detected but causes most of the damage.

Two types of surface-coating materials should be looked at. The waterproofing agents are impervious coatings, such as coal tar, asphalt, and some paints, which seal the surface to liquid water and water vapor, and normally are intended for below-grade use (foundations, basement walls, etc. The water repellents, materials for above-grade use, allow the masonry to breathe. That is, they permit the movement of water vapor, but no infiltration of liquid water, although pressure greater than a 2-in. head of water can cause penetration. An impermeable surface film above grade not only would prevent the entrance of water into the masonry, but also would prevent the loss of excess water from its interior.

REFERENCE

Bryant, Terry, "Protecting Exterior Masonry from Water Damage," *Technology and Preservation Magazine,* Spring 1978, Vol. 3, No. 1.

A Caution—Sealers

The indiscriminate application of silicones and other clear penetrating sealer solution to brick masonry can often cause more harm than good, warns the Brick Institute of America.

Among the possible dangers are:

1. It will not stop moisture penetration through cracks or incompletely filled joints.
2. It could cause or contribute to spalling and/or disintegration of the brick units, by causing crystalline deposits or salts to form within the unit.
3. It will not completely stop staining and efflorescence, and may cover it sufficiently to prevent its removal.
4. It could make the wall nearly impossible to tuck point, if that were required.

The general recommendation of the Brick Institute, based on many years of experience and study, is not to use silicones or other clear penetrating solutions of similar ilk. If they must be used; they should be applied only in specifically controlled and selected circumstances.

Flashing

Flashing provides a degree of control over moisture penetration and movement within masonry. It is difficult to completely prevent rainwater from entering at parapets, sills, projections, recesses, roof intersections, etc. unless proper flashing is installed. Without proper flashing, water which does penetrate a wall will not be diverted back to the exterior of the building but will flow inward.

When investigating a masonry structure for water infiltration certain basic practices in installing flashing can be checked. When flashing extends to the interior its end should have been placed between furring and the interior finish and turned up at least 1 in. to collect moisture that may penetrate through the wall.

Any moisture which does enter a wall gradually travels downward. Flashing should have been placed above grade at the wall base to divert this water to the exterior. Upward capillary travel of ground moisture, especially where there are no basements, should have had damp checks placed about 6 in. above grade.

Windowsills should have had flashing placed under and behind them except perhaps where impervious monolithic sills have been used. The ends of sill flashing should extend beyond the jamb line on both sides and be turned up at least 1 in. into the wall slope. All sills should, of course, drain water away from the building.

All flashing should be drained to the outside. Tooled mortar joints do not drain without weep holes.

If the building has been intelligently flashed the corrosive effect of fresh mortar on flashing should not be overlooked.

Zinc. Galvanized coatings are subject to corrosion in fresh mortar. Although the corrosion products apparently form a very compact film around zinc and apparently form a very good bond with the mortar, the extent of corrosion cannot be accurately predicted. Some zinc-alloy flashings are available. Like many alloys these may have properties considerably different from those of the pure metal.

Aluminum. The caustic alkalies present in fresh, unhardened mortar will attack aluminum. Although dry, seasoned mortar will not affect aluminum, corrosion can occur if the adjacent mortar becomes wet. Since the purpose of flashing is to direct the flow of water, it is apparent that uncoated aluminum will prove unsatisfactory in flashing.

Lead. Lead, like aluminum, is susceptible to corrosion in fresh mortar. Furthermore, when lead is only partially imbedded in mortar, in the presence of moisture it develops a different electric potential, acting as the positive element of an electric cell. The resulting electrolytic action gradually disintegrates the imbedded lead.

Plastics. These are widely used. The best are tough, resilient, and highly resistant to corrosion. However, because the chemical compositions of plastics are so highly diversified it is impossible to lump all plastic flashings into a generalized group. Some will not withstand the corrosive effects of masonry mortars. The investigation must rely on performance records of the material, the reputation of the manufacturer if this is possible, and perhaps laboratory analysis.

REFERENCE

Brick Institute of America, *Technical Notes.*

EFFLORESCENCE

Efflorescence is a deposit of water-soluble salts upon the surface or in the pores of masonry. It is usually white in color, although certain vanadium and molybdenum compounds, present in some ceramic units, produce a green deposit usually referred to as *green stain*. There is also an occasional instance of *brown stain* resulting from deposits of manganese compounds.

The principal objections to efflorescence are usually based on the appearance characteristics. However, while surface efflorescence is unsightly and a nuisance to remove, it is by no means as harmful as the efflorescent crystal pressure resulting from deposits within the pores of the masonry. When salts are deposited

Efflorescence at window head. (Photo by F. Wilson)

Efflorescence at window head. (Photo by F. Wilson)

below the surface of brick in the pores the force of crystallization and growth of crystals can cause cracking and disintegration of the material.

The mechanisms of efflorescence are complicated, but they can be simply stated as water-soluble salts brought to the surface of the masonry in solutions of water and deposited there by evaporation. Salts may migrate across the surfaces of the units or through the pore structure of the masonry.

Certain conditions must occur simultaneously for salts to appear. Soluble salts must be present within or in contact with the masonry assembly. They may be present in the facing units, backup, mortar ingredients, or trim. There must be a source of water and it must be in contact with the salts for sufficient time to move them into solution. The construction assembly must be such that pathways exist for the migration of the salt solution to the surface of the masonry, or other locations, where the salts are deposited by evaporation of the water.

If masonry could be constructed without water or if no water penetrated the masonry, even where water soluble salts exist no efflorescence would occur. These conditions, however, cannot be met.

Sources of Salts

The chemical composition of efflorescence salts is most commonly alkali and alkaline earth sulfates, and carbonates, although chlorides have also been identified on rare occasions. The most common salts found in efflorescence are sulfate and carbonate compounds of sodium, potassium, calcium, magnesium, and aluminum. In rare instances, chlorides may occur as efflorescence, but this is usually a result of contamination of masonry units or mortar sand by sea water or the runoff from alkaline soils.

The problem of locating the source of efflorescence is complicated by the many available sources of soluble salts. They may be present in the masonry units or in the mortar, or they may result from the penetration of the masonry by rainwater or ground water.

Simply because efflorescence appears on the face of the brick it is usually assumed that the brick is at fault, but this is not normally the case. However, due to the composition of the raw materials and the heat treatment used in manufacture, it is possible for soluble salts to appear in the finished product. If water is absorbed by such products the soluble salts enter into solution and efflorescence results as evaporation takes place. But bricks such as these will show efflorescence even if placed in distilled water and all precautions taken to eliminate outside contamination.

Many materials used as backup for brick masonry walls contain large quantities of soluble salts. If sufficient water is present to dissolve the salts and pathways are provided for the solution to reach the surface of the masonry the salts will appear there. A test by J. E. Young showed that concrete products contain two to seven times as much soluble material as brick.

Trim material such as caps, coping, sills, lintels, keystones of natural stone, caststone, and precast concrete, may contain soluble salts and transfer it to adjacent brickwork.

A primary and obvious source of contamination is the mortar used in wall construction. It is in intimate contact with the brick on at least four and sometimes five sides. It is applied to the brick in a wet, pastelike condition which provides ample moisture for the transfer of soluble salts from the mortar to the brick. If salts are available in the mortar it will be transferred to the brick.

Experience and laboratory data indicate that lime is seldom a contributor to efflorescence. Sands are primarily silica and are not water soluble. However, they may themselves be contaminated with material that will contribute to efflorescence, such as sea water, soil runoff, plant life, and decomposed organic compounds.

In addition to rainwater and ground water which may penetrate masonry walls, water may accumulate within the wall as a result of condensation of water vapor. Frequently, efflorescence that appears on rain-resistant masonry walls is due to this accumulation of water.

Condensation is usually due to moisture originating inside the buildings. The moisture content of the outside air, which enters the building and is heated for comfort purposes, is invariably increased by moisture released from cooking, bathing, washing, and other operations employing water or steam, and by the moisture released by exhalation and perspiration from the occupants. This gain in moisture content of the air increases the vapor pressure substantially above that existing in the outdoor atmosphere. This increased pressure tends to drive the vapor outward from the building through any vaporous materials that may comprise the enclosing surfaces.

When vapor passes through porous, homogeneous materials, which may be warm on one side and cold on the other, it may pass through the zone of its dew point temperature without condensing into water. But if the flow of vapor is impeded by vapor-resistant surfaces at a temperature below the dew point temperatures the vapor may condense upon such cold surfaces. This condensed moisture can contribute to efflorescence on the surface of the wall.

Analysis Procedure

An examination with the following checklist may be sufficient to determine the cause and extent of the problem, and suggest methods for its repair and alleviation.

1. Determine the age of the structure at the time when the salts first appeared. If "new-building bloom" is involved (structures less than one year old), the source of the salts is often the cement in the mortar, and the source of the moisture is usually the construction water. If, however, the building is over a year old, other sources must be considered. If the structure showing efflorescence is over two years old, construction details should be examined for possible leaks in the wall or in the surrounding construction. The appearance of efflorescence on an established building, which has been free of efflorescence, is usually attributable to a new entry of water into the construction assembly. It is assumed in this instance that the soluble salts are present in the construction and that the addition of moisture makes them appear.

2. The location of the efflorescence, both on the structure and on the individual units or mortar joints, should also be carefully noted. The location on the building may offer some information as to where the water is entering. The location of the salt crystals on the joints or the units may be of help in determining the source of the salts. The recent use or occupancy of the building may also be noted. For example, has it been vacant for some time, has new construction been going on? In short, what has changed that might cause or trigger the appearance of the efflorescence?

3. The condition of the masonry should be carefully examined. The profile of the mortar joints, the condition of the mortar, the type of workmanship which was employed, the condition of caulking and sealant joints, the condition of flashing and drips, any deterioration or eroding of mortar joints in copings or in sills, all should be carefully noted. This information should offer clues as to the entry paths of moisture into the construction.

4. The wall section and details of construction should be examined for an indication of possible paths of moisture travel, and for possible sources of contamination by soluble salts. A careful examination of roof and wall juncture and flashing details should be made. A comparison of contract drawings with as-built drawings may also be helpful. This examination will also be useful for the later determination of steps for repair to eliminate the efflorescence.

5. Laboratory test reports on the materials of construction should be examined, if they are available. This will help determine the source of the soluble salts, and may be of use in analyzing and making repair judgments.

6. The identification of efflorescence salts is sometimes of use. This can be done by commercial testing laboratories. X-ray diffraction analysis is sometimes used. Petrographic analysis or chemical analysis is also possible. In some instances it is useful to know both the type of salts present and their relative quantity. According to Brownell, "It is most unusual to find two or more salts comprising the efflorescence. . . . The presence of carbonate indicates mortar contamination, soluble sulfates place the brick in suspect, and potassium chloride most certainly is the result of cleaning the structure with muriatic acid. Sodium chloride efflorescence may very well be traced to sea water, either from the sea itself, mist in the air, or absorbed up from the ground."

7. Miscellaneous sources of water should also be considered, if all other sources seem to be eliminated. Some of these sources are: condensation within the wall, leaky pipes, faulty drains, and condensation on heating or plumbing pipes.

CLEANING BRICKWORK

Masonry walls are usually cleaned several times during the life of a building; the first cleaning occurs even before the building goes into service. The general dirt and dust of construction, including mortar splashed or smeared on the bricks as a result of the bricklaying procedure have to be removed.

A second cleaning may be required to remove efflorescence. Because mortar must be wet to render it plastic for bricklaying, brickwork becomes damp in the course of construction. As it subsequently dries, the moisture, in which the various salts derived from mortar and bricks have been dissolved, moves to the wall surface to evaporate, leaving a deposit of salts, usually as a white coating on the bricks.

The next cleaning of the brick walls will depend upon the amount of soiling they receive in service. This is governed by the nature of the atmosphere to which they are exposed.

The chief culprit in the staining and dirtying of buildings in the past has been the burning of coal. Coal burning releases to the air materials such as soot; tarry matter; dust particles; various vapors and gases including carbon monoxide and carbon dioxide, water vapor, sulfur dioxide, and oxides of nitrogen; as well as organic compounds, particularly hydrocarbons. The burning of other fuels such as oil and natural gas similarly produces soot; vapors and gases, including sulfur dioxide, ammonia, methane, and acetylene; plus organic materials.

The particles of matter produced by the combustion of fuels and by such industrial processes as the heat treatment of materials may remain suspended in the air for a long time before settling out if, in the meantime, they have not collided with and attached themselves to a wall. The considerable amount of such matter in the atmosphere is illustrated by dustfall collections taken in the industrial area of Windsor, Ontario. Here, average dustfall during the heating season is about 92 tons per square mile per month, with peak values as high as

(Photo by F. Wilson)

200 tons. In contrast, natural dustfall; wind-blown sand, soil particles, pollen and vegetation, probably amounts to about 5 tons per square mile per month.

The need to clean brickwork may arise from many causes, including atmospheric pollution such as stains from metals, particularly iron and copper used in conjunction with brickwork, and growth of lichens, moss, and other vegetation.

The materials employed in cleaning brickwork are highly corrosive and frequently toxic. They require special equipment and specialists to handle them, protective clothing, and protection of the surrounding environment.

The bricks of most buildings are made of burned clay or shale, but many structures have been built of sand-lime (calcium silicate) and concrete bricks. Acid cleaning solutions that have no effect on clay bricks may harm the others, and for these the strength of the cleaning solution must be reduced.

REFERENCE

Richie, T., *Cleaning of Brickwork,* Canadian Building Digest Division of Building Research, National Research Council of Canada, April 1978 CBD 194.

REPOINTING MORTAR JOINTS

Damp walls may be caused by leaks from parapets, flashing, or roofs, and such leaks may appear some distance below in the masonry walls. Another source is rising capillary moisture, which can cause dampness several feet above the ground. In either of these cases, repointing the outer masonry wall will not cure the problem. Similarly, if open joints or loose bricks are caused by foundation settlement or deterioration of materials, such structural problems should be corrected before beginning masonry work.

Repointing is an expensive and time-consuming task because of the large amount of hand work and special materials required. Hand-made brick may have a long delivery time. Scaffolding may be in place for an extended period, possibly interfering with normal circulation patterns.

If water penetration is to be corrected, the repointing work probably should come early in the preservation project; if the building is watertight, it may be better to wait until after completion of exterior cleaning so that the existing and new mortars will weather simultaneously.

Repointing may require analysis of the mortar, the bricks, and the techniques originally used in striking the joints.

Identifying Original Mortar in Color, Texture, and Other Properties

The identification of the major constituents is relatively simple. It may be possible to arrive at an approximation of the mortar's composition by observation, using low magnification of an unweathered sample. Once the constituents have been separated and identified, the sand can be examined for the range of color, size, and shape of the grains. Other insoluble materials should be identified for inclusion in the repointing mortar.

It is a common error to assume that hardness or high strength is a measure of

Cracks in mortar joint. University of Maryland. (Photo by F. Wilson)

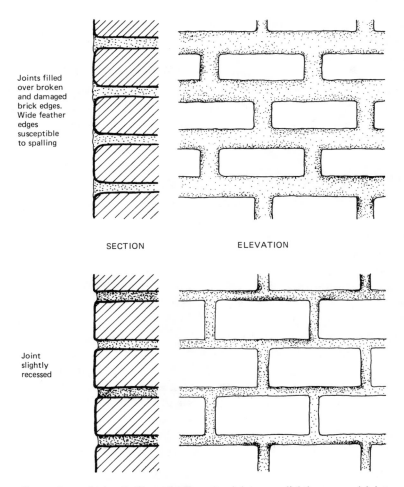

Joints filled over broken and damaged brick edges. Wide feather edges susceptible to spalling

SECTION

ELEVATION

Joint slightly recessed

Comparison of visual effect of full mortar joints vs. slightly recessed joints. Filling joints too full hides the actual joint thickness and changes the character of the original brickwork. (After Preservation Briefs #2-National Park Service, Lee Nelson, AIA, Editor)

durability. Stresses within a wall caused by expansion and contraction or by settlement must be accommodated in some manner; in a masonry wall these stresses should be relieved by the mortar rather than the brick. A mortar which is stronger and harder than the masonry units will not give, thus causing the stresses to be relieved by the masonry units, usually in the form of cracking and spalling. Uneven movement in the masonry also can break the bond between the mortar and the brick, opening hairline cracks to water penetration. Mortars with a high percentage of Portland cement can have the above described deleterious effects. Porous mortar permits water within the wall to migrate and escape. Mortar with a high cement content does not permit this movement, and the water trapped within the wall may be subjected to freeze-thaw cycles which can spall soft, older brick.

"Workability" or plasticity of the mortar also is important. The new mortar should have both cohesive and adhesive qualities to make complete physical contact with the masonry and old mortar.

It should have the maximum amount of sand consistent with such workability to help reduce shrinkage while drying. The mortar must not be sticky or gummy, and must handle readily on the pointing tool. Finally, the newly applied mortar must have good water retention to resist rapid loss of water through absorption by the brick or old mortar while setting.

Advantages of Using High-Lime Mortar

These facts lead to the conclusion that a high-lime mortar generally is best for most historic structures, even those originally constructed with cement mortars. High-lime mortar is soft and porous, and has the lowest volume change due to climatic condition. In addition, lime mortar is slightly soluble in water and able to self-seal small cracks and voids that may develop. In this phenomenon, a slight amount of the mortar dissolves in rainwater and is precipitated in the void during the drying process, thus sealing the crack. Even straight lime mortar is more durable than generally recognized as long as the wall is protected from water penetration with sound roofing, gutters, flashing, etc. A small amount of white Portland cement may be desirable, however, to accelerate setting.

Even if the building originally was constructed with cement mortar, it usually is best to use a high lime mortar rather than match the original. High lime mortar will reduce potential stresses at the edges of the masonry and also help minimize shrinkage, which leads to hairline cracking.

Joint Preparation. Generally old mortar should be cut to a minimum depth of 1 in. to insure an adequate bond between the new mortar and the existing masonry and to prevent mortar "popouts." For joints less than ⅜ n. thick, cutting the mortar back ½ in. usually is sufficient if the mortar behind that point is in good condition. Any loose or disintegrated mortar beyond this minimum depth should be removed. Unless the mason is unusually skilled and extremely careful, the use of power tools for mortar removal inevitably will damage the brick. Damage to the edges of the brick will significantly affect the character of the brickwork; in addition, absorption of water is increased since the softer inside of the brick is no longer protected by the hard burned outer surface. Where joints are uniform and fairly wide, and the bricks were machine-made with straight edges, it may be possible to use a grinder. A test patch will establish the feasibility of using a grinder. If there is any chance of damage to the masonry occurring, however, hand methods should be used exclusively. Although they are slower, they are easier to control, and less likely to cause irreversible damage.

REFERENCE

Mack, Robert C., *Preservation Briefs: Repointing Mortar Joints in Historic Brick Buildings,* #2, Technical Preservation Services Division, Office of Archeology and Historic Preservation/Heritage Conservation and Recreation Service.

EVALUATING MASONRY—TEST METHODS

Since many existing buildings being considered for rehabilitation are constructed of masonry, it is important to know how to evaluate the condition of the masonry and to determine whether it is capable of supporting the loads imposed by the retrofit or rehabilitation work. This is especially true of load-bearing masonry assemblages which are expected to support new heavy dead or live loads.

Several nondestructive evaluation methods have been used with various degrees of success to determine the physical properties of masonry units and mortar. However, these test methods have limited application and generally provide information only on the physical make-up of the masonry (continuity, location of voids, reinforcement, etc.). In only one case (low-frequency ultrasonics) can an estimate of compressive strength be obtained (by experienced operators and evaluators), but it is prohibitively expensive to use in routine investigations. One manufacturer claims to be making progress in the development of a penetration device known as the Densicon Penetrometer which is based on the same principle as the Windsor Probe for concrete evaluation, and it might become an accepted test for measuring surface hardness (from which it may be possible to estimate strength). As of this time, definite conclusions concerning its effectiveness cannot be drawn by the authors.

In contrast to the limited choice and undetermined reliability of nondestructive methods for evaluating in-situ strength properties of masonry, there are a large number of more thoroughly developed methods available for this purpose which are termed destructive or which require testing other than in-situ. Depending on the user's need for information, the practicality of performing these tests, funding availability, etc., these may be particularly applicable in rehabilitation projects.

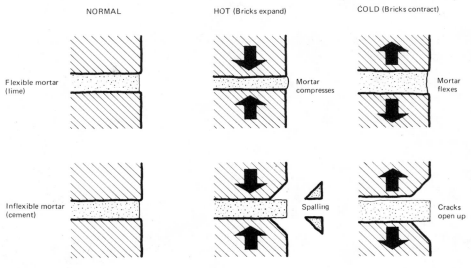

Diagramatic sketches showing effects of temperature change upon masonry. Flexible mortor expands and contracts with temperature changes. Bricks bonded by inflexible mortar tend to spall at the edges, the area of greatest stress, in hot weather and seperate from the mortar when it is cold. This latter condition opens cracks, permitting the entry of water and causing additional deterioration. (Preservation Briefs #2, National Park Service, Lee H. Nelson, AIA)

To allow the reader the opportunity of selecting the most appropriate test, the accompanying table shows a reasonably complete listing, and summary description of the commonly used nondestructive and destructive test methods available to evaluate masonry. Test methods are grouped in the table according to the parameter being tested.

Test Methods for Masonry

MATERIAL	TEST PARAMETER	TEST METHOD	COMMENTS
Masonry assemblages (units and mortar)	Flexural bond strength: Brick, sampled and tested per ASTM C 67; Concrete, sampled and tested per ASTM C 140; Mortar, selected and tested per ASTM C 270 and ASTM C 518	Use either the third-point loading method or air bag. Specimen is loaded as a simply supported beam and is placed horizontally on its supports in the testing machine	Testing machine must conform to requirements of Method E 4
	Diagonal tensile or shear strength.	Masonry assemblages are tested by loading them in compression along one diagonal, thus causing a diagonal tension failure with the specimen splitting apart parallel to the direction of load	Use a 1.2 × 1.2 m (4 × 4 ft) masonry assemblage. This method eliminates the need for a hold-down force (to prevent rotation of the test specimen) as required in the racking load test prescribed in Methods E 72. Three test specimens should be used
	Water absorption; saturation coefficients (for prediction of durability).	Weighing dry and saturated conditions of test sample. Sample is saturated by submersion in boiling water for 5 hours and cold water for 24 hours	Scale should have capacity of at least 2000 g, and should have a sensitivity of at least 0.5 g for brick or 0.2% of the weight of the smallest specimen for tile
	Freezing and thawing, resistance to damage	Repetitive cycles of wetting, freezing, drying, and weighing	Test is continued through 50 cycles of freezing and thawing unless specimen breaks or loses more than 3% of its original weight as judged by visual inspection
Masonry	Size	Visual measurement	Use either a steel metric (or 1 ft) scale, or a gauge or caliper with a scale ranging from 25 to 300 mm (1 to 12 in) and having parallel jaws. Of no value in determining strength or durability
	Warpage	Visual measurement using a scale or measuring wedge	Use either a steel metric (or 1 ft) scale or a steel measuring wedge. The wedge shall be numbered to show the thickness of the units. Of no value in determinnig strength or durability

Ceramic glazed facing tile and brick	Imperviousness	Permanent blue-black fountain pen ink is applied to the glazed surface of 5 dry specimens for 5 minutes. The surface is washed and examined for stain of the finish	Of no value in determining strength or durability
	Chemical resistance	End portion of test specimen is dipped in 1½ in. of a 10% solution of HCL for 3 hours. The opposite end is dipped in a 10% solution of KOH for 3 hours. Finishes are then rinsed, dried, and examined visually for changes in texture or color	The solution must be maintained at 15–20°C (60–80°F) temperature. Of no value in determining strength. Could be useful if certain chemicals are anticipated to come in contact with the masonry
Ceramic glazed structural clay facing tile, facing brick, and solid masonry units	Crazing	Autoclave crazing test. Test specimens are placed in an autoclave with 150 pst steam pressure for 2–2½ hours. Specimens then are cooled slowly to room temperature. Then permanent blue-black fountain pen ink is applied to the glazed surfaces and a visual inspection is made to detect crazing	Cooling should be extended over a minimum of 3 hours. Normal safety precautions should be observed concerning autoclave operation. Of no value in determining strength or durability
	Opacity	Permanent blue-black fountain pen ink is applied to the test specimens along a 50-mm (2-in.) length of the edge of the finished surface. After 5 minutes, the finish is visually examined for opacity	Of no value in determining strength or durability
Unit masonry	Leakage (water permeance)	The masonry specimen is placed in a spray test chamber which has controlled air pressure. Streams of water impringe against the exposed surface of the specimen at a rate 139 l/m² (3.4 gal/ft²) per hour. The air pressure is raised to 479 Pa (10 lbf/ft²) above atmospheric pressure. The specimen is then dried and the back (unexposed) face is painted with a thin coating of whitewash. The test is repeated (after the whitewash dries for a minumum of 24 hours). This final test is conducted for 3 days or until a rating has been attained	The time for appearance of visible water on the specimen is observed. The rate of leakage is observed; and from this, the water permeance is rated in accordance with ASTM E 514 classification (Class E, G, F, P, or L). This test is used mainly for a comparison of masonry specimens. It is not good for an acceptance/rejection test. Simple modifications make it useful for testing of in-situ masonry walls

Portland cement–lime–sand mortar and masonry cement–sand mortar	Compressive strength, water retention	Standard compression tests and water retention tests are used in accordance with ASTM C 91 with exceptions per BIA MI-72	Types M, S, N, and O are included. Use type I, II, or III Portland Cement per ASTM C 150; hydrated Lime per ASTM C 207; sand per ASTM C 144. No air entrainment or antifreeze admixtures shall be used. Mixing and proportioning per BIA MI-72
	Air content	Air content is determined in accordance with ASTM C 231	
	Efflorescence	Efflorescence tendency is determined using the wick test described in BIA Research Report No. 15, Sec. 4.4, p. 14	
Masonry (face brick, sandlime brick, concrete block, structural clay tile, and mortar) including the full assemblage	Compressive strength	Method B, (ASTM E 447), methods of test for compressive strength of masonry prisms is used to determine compressive strength of existing masonry built with the same materials as used in the test sample. A minimum of three test prisms are built with like materials (same as in-situ). No reinforcement is used (except metal ties). Compressive strength is determined from 7-day and 28-day tests, and Young's modulus can be determined in accordance with Method E III	Test apparatus must conform to requirements of Method E 4. Test of building and face brick is in accordance with ASTM C67; sandlime brick, C 67; concrete block, C 140; structural clay tile, C 67; mortar, method C 109. Reliability of test results is uncertain because of the unlikelihood that test materials are exactly the same as in-situ materials
	Structural soundness of units, bond with mortar, and to determine if cells are filled	Hammer test: Lightly tap the masonry unit with a hammer. Listen to resonant sound. A very experienced evaluator might be able to determine the condition by the sound	This test requires an experienced person with a good sense of hearing and a delicate touch. It is an unsophisticated test with questionable results. Test scores may be needed to validate findings
	Location and uniformity of the inner cell grout and wall thickness.	Probe holes: Penetrate the area of investigation with a small masonry bit and probe the hole with a stiff wire	Small holes may be patched easily. Surface damage is only minor

Masonry units and mortar (including the assemblages)	Continuity (voids or cracks), and estimation of compressive strength	Low-frequency ultrasonics: Soniscope and two transducers are used. Transmitter and receiver are placed on opposite sides of the masonry unit. Low-frequency ultrasonic sound waves are transmitted. Travel time and relative strength of transmitted signals is measured. Voids or cracks in units will weaken the signal. Compressive strength can be estimated by correlation of pulse velocity through units and mortar with compressive strength of cores or prism that were removed from the wall tested	This equipment usually is available only through specialized consultants and requires operators and evaluators who are very experienced with testing of masonry. The cost may be prohibitive for routine investigations
	Location of voids and/or reinforcement.	Gamma radiography: The gamma source and x-ray film are placed on opposite sides of the test specimen. After exposure of the film for several minutes, the film is processed and read. Voids show on the film as dark irregular patches. Reinforcement shows as a light area on the film	Access to two sides of the test specimen is required. Extensive safety procedures are required due to the health hazard of gamma ray exposure. The cost could be prohibitive for routine projects. It is used mainly for specialized cases such as distressed precast masonry units when a record is needed for possible litigation
	Location of steel reinforcement	Pachometer: The pachometer is a magnetic detector. The operation is based on the principle that a ferromagnetic component will cause a variation in the magnetic field induced into the masonry (a nonmagnetic medium)	The surface of the masonry unit is scanned with a probe. Readings indicate location, size, and depth of reinforcement. This test is used only for light reinforcement. If both joint and cell reinforcement are used, results are difficult to interpret.

REFERENCE

Selected Methods for Condition Assessment of Structural, HVAC Plumbing and Electrical Systems in Existing Buildings; NBSIR 80-2171. Frank H. Lerchen, James H. Pielert, Thomas K. Faison.

EVALUATING MASONRY—A Reminder

Four components must be considered for "unit masonry" materials: the units themselves, the mortar, ties, and reinforcements. Although modern units are produced under highly controlled conditions, variations in strength, durability, etc., can be expected for older products. This is also true for mortars, which until fairly recently were lime-based and did not include Portland cement. The compressive strengths of mortars vary in the extreme variable from less than 100 psi to over 3000 psi. Brick units, depending on their raw materials, firing, cutting, moisture, etc. can vary from less than 1500 psi to more than 16,000 psi in compressive strength. Structural tiles, which were a very popular construction from about 1900 to 1950 have similar variations. Compressive strengths (net area, edge tested) vary from 1100 psi to over 10,000 psi.

Concrete block, or concrete masonry unit, or cmu, manufacture has become a highly controlled system in the last few decades, but prior to this small operations produced blocks of grossly varying characteristics. The compressive values to be expected for blocks of this type, which can often be recognized by irregular texture, voids, out-of-squareness and imprecise edges, can be from less than 100 psi net section to over 1000 psi. Contemporary quality-controlled units, in contrast, can be expected to be at least 1000 psi to over 3000 psi.

REFERENCE

Fitzsimmons, Neal, C. E., *Research Support for Building Rehabilitation Studies in the Area of Strength and Stability Evaluation*, National Bureau of Standards, 1979.

PHYSICAL PROPERTIES OF CEMENT MASONRY

The physical properties of concrete masonry units are determined largely by the properties of the hardened cement paste and the aggregate. Much of the technical knowledge relating to concrete is applicable to concrete brick and block.

Differences occur in mix composition and consistency, consolidation method, textural requirements, method of curing, and other factors. Compared with structural concrete, concrete masonry units generally are made with a substantially lower cement factor (3–4 sacks of cement per cu. yd, compared to 4–7 in concrete) and a much lower water–cement ratio (2–4 gallons per sack of cement versus 5–7). Aggregate is graded finer, with the largest size seldom in excess of ⅜". Concrete blocks are also often made of porous or lightweight materials. Concrete masonry units contain a relatively large volume of interparticle void spaces (spaces between aggregate particles not filled with cement paste), and are cured at much higher temperatures.

The compressive strength of concrete masonry units is difficult to predict from mix data, since a simple water–cement ratio is not valid for harsh, dry mixes, and because each type of aggregate exhibits different characteristics during mixing. The highest strengths are obtained from the wettest moldable mixes, but generally the maximum amount of mixing water is not used in order to cut down breakage of the freshly molded units during the handling operations prior to curing. Drier mixes do not compact as well as wetter ones and therefore have lower strengths. Each manufacturer experiments to develop mixes and techniques to produce units of highest strengths compatible with other desirable handling properties.

Solid face and veneer units such as concrete brick and split block, when made with sand and gravel or crushed stone aggregate, will generally exceed 3,000 psi in compression. The principal factors influencing compressive strength are (1) type and graduation of aggregate, (2) type and amount of cementitious material, (3) the degree of compaction attained in molding the concrete units, and (4) moisture content and temperature of the units at the time of testing. Units made with lightweight aggregates generally exhibit lower compressive strengths than units made with heavier sand and gravel, limestone, and air-cooled slags.

Tensile strength, flexural strength and modulus of elasticity of concrete masonry vary in proportion to compressive strength. Tensile strength will range generally from 7 to 10% of the compressive strength, flexural strength from 15 to 20% of compressive strength, and the modulus of elasticity from 300 to 1200 times compressive strength.

Testing

Water absorption tests can provide a measure of the density of the concrete. Absorption is a measure of the pounds of water per cu. ft of concrete and in hollow concrete block varies over a wide range, from as little as 4–5 lb per cu. ft for the heaviest sand and gravel units to as much as 20 lb per cu. ft for the very porous, lightweight aggregate types. Solid units made with sand and gravel will normally have less than 7 lb per cu ft absorption. The absorption of the aggregate, which may vary from 1–5% of the dry weight of aggregate for dense aggregates to as much as 30% or more for the lightest aggregate, has strong influence on the absorption of the units.

The porosity and pore structure of the concrete influence other properties, such as permeability, thermal conductivity and sound absorption, but the influence is not always predictable, especially when comparing units made with different types of aggregates. Higher water absorption is not a desirable or purposely incorporated property, but it is accepted as a natural consequence when porosity properties such as lightness in weight, higher sound absorption, or thermal insulation are desirable.

A high initial rate of absorption or suction rate results when concrete contains a large portion of relatively large, interconnected pores and voids. Units with a high suction rate combined with high absorption will have high permeability to water, air, and sound and may also have less resistance to frost action. On the other hand, unconnected, air-filled pores present in lightweight aggregate and in air-entrained cement paste impart the advantages due to porosity yet minimize permeability to water, air, and sound.

An abnormally high suction rate adversely affects the structural bond of mortar to units but can be minimized by using mortars with high water retention properties.

All block intended for use in exterior walls and which will not be painted should have low absorption, and mortar joints should be carefully tooled for weathertightness.

Volume Changes Due to Moisture

Concrete masonry units undergo small dimensional variations due to changes in temperature, changes in moisture content, and a chemical reaction called carbonation. Of these factors, only moisture changes can be conveniently minimized to a significant degree. This can be done through preshrinking the units by adequate drying before they are built into the wall.

Carbonation causes irreversible shrinkage in the units as a result of a chemical reaction within the concrete when it absorbs carbon dioxide from the air. Though little test data exists, it appears that the magnitude of the change under certain conditions and over extended periods of time may approximate that due to moisture-content change.

Temperature changes cause units to expand when heated and contract when cooling. The volume changes are fully reversible, causing the unit to return to its original length after being heated and cooled through the same temperature range. The coefficient of thermal expansion for concrete block is dependent largely upon the coefficient of the aggregate, since it comprises roughly 80% or more of the concrete volume.

Moisture changes cause units to expand when wetted and contract when drying. Moisture volume changes are not quite fully reversible, since during the first few cycles of wetting and drying there is a tendency for concrete to assume a permanent contraction. However, in subsequent cycles the movement is reversible.

An important factor contributing to the development of cracks in concrete masonry walls is the volume change in the units due to original drying shrinkage. If the units are laid up in the wall before they have become dimensionally stable and allowed to shrink, tensile stresses will develop wherever the wall is restrained from shrinking, and cracking can be expected. Drying shrinkage can be greatly reduced with proper curing and drying so that the moisture content of the units when they are placed in the wall is in equilibrium with the surrounding air; that is, the units are dried down to the average air-dry condition to which the finished wall will be exposed in service.

The amount of residual shrinkage, or movement, a concrete unit will undergo due to moisture change after placement in a wall is the product of its shrinkage potential and its moisture loss after placement (Residual shrinkage = shrinkage potential × moisture loss.) Shrinkage potential depends upon manufacturing technique, raw materials, age of block and other factors. Moisture loss depends upon the moisture content of the units when laid in the wall and the relative humidity and temperature of the drying environment. While measurable in the laboratory, the small magnitude of the phenomenon makes it impractical to measure shrinkage on a routine basis as a means of demonstrating compliance with specifications.

Principal benefits of high-pressure steam curing are high early strength and greater stabilization against volume changes caused by varying moisture conditions. Units can be produced for use withim 24 hours after molding which develop permanent compressive strength at one day, equal to that of moist-cured units at 28 days. Concrete cured in an autoclave undergoes a different chemical reaction during hydration than concrete cured at atmospheric pressure, and produces more stable units with less tendency to shrink from drying or expand from wetting. The shrinkage of high-pressure, steam-cured units in drying from a saturated condition to equilibrium with air in a heated building is about 50% less than similar moist-cured units.

Concrete unit masonry construction, like brick masonry, is quite rigid and readily affected by building movement. Cracking and movement problems arising from drying shrinkage are not usually serious if the blocks exhibit no more than 0.05% shrinkage from saturated to dry condition. Blocks made of lightweight aggregate may show shrinkage up to 0.08% or even higher and may exhibit serious cracking problems unless used in such a way as to control the moisture changes subsequent to laying. Masonry made from concrete bricks may also be subject to cracking.

REFERENCE

Olin, Schmidt, Lewis: *Construction: Principles, Materials and Methods,* Section 205—Masonry, The Institute of Financial Education, Chicago Illinois, and Interstate Printers and Publishers, Danville, Illinois, 4th Edition, 1980.

STONE

The factors that affect the deterioration of stone include its composition, structure, and surface conditions, interaction of the surrounding microclimate and environment with the stone and the methods that temper these destructive tendencies.

Dimension stone is among the most durable of materials, but the evidences of weathering indicates its ultimate vulnerability. The Committee on Conservation of Historic Stone Buildings and Monuments, of the National Research Council of the National Academy of Sciences, asserts in their recent authoritative report that there is much too little information concerning stone's vulnerability. They note that stones are usually selected because of its aesthetic qualities, with casual reference, at best, to basic data on porosity, pore size, moisture absorption, and other critical physical and chemical parameters. A more effective exchange of information on stone, based on its actual performance in place, rather than the development of new laboratory information were among the committee's recommendations.

Information, they claimed, is needed on the variability of stone formations, of stone within a quarry, and within the rock formation itself. Most rock formations are not homogeneous. One cannot simply take a sample of rock from a quarry, test it, and extrapolate the results to the entire quarry or formation.

In addition to observing weathering history in the field it must be determined if weathering can be simulated in the laboratory. The Committee on the Conservation of Historic Stone Buildings and Monuments is not optimistic that this can be done.

Natural outcroppings of building stones have a quite different environmental exposure than those stones set in buildings. In any stone structure a humidity gradient occurs from the outside in. If a building is air-conditioned in the summer and heated in the winter it experiences constantly changing humidity cycles. Natural stone outcroppings are not exposed to water by capillary action as extensively as are building stones nor are they sprayed with salt to remove snow and ice. A building is much more than a small sample of stone in a laboratory. It is a dynamic system, consisting of a variety of materials reacting to a range of stresses. It must be understood how each factor relates to the whole with weathering and decay studied on a systematic basis.

The committee also found that there was not adequate literature available to describe stone decay. There is some information on some aspects of the decay problem but nearly all facets of this phenomenon require further sustained study. Perhaps the most important matter that must be thoroughly understood is the effect of moisture as part of the freeze-thaw cycle and water as a vehicle for migrating salts. Another related problem is the structure of rocks themselves. Stress-strain relationships are nonlinear. Microcracks not visible to the naked eye are extremely important factors in the stone's weathering. These might provide access for movement of dilute solutions in rocks and so accelerate the spalling phenomena associated with damage by salts.

The committee also drew attention to the fact that more information must be developed on wet and dry deposition of air pollutants and on the entire spectrum of pollutants, not only sulfates and nitrates. Very little is known of air quality and microclimate in terms of the management of stone structures. Considerable difficulty is experienced in attempting to specify the range of monitoring required for buildings. It is not possible today to relate data accumulated at established air-monitoring stations to building sites that are often distant from these stations.

The roles of vibration caused by vehicles and of biological agents in the destruction of stone are other aspects of deterioration that need to be better understood. We need, the committee concluded, to develop a general theory of the decay of stone that includes all appropriate anthropogenic and natural phenomena. This must include the effects of freeze-thaw cycles. Wet and dry deposition of air pollutants, biological attacks, and damage by salts.

Diagnosis of Deterioration

In addition to their own weight, stone structures must often sustain such forces as geophysical loads from wind, temperature, and earth tremors; geotechnical loads from lateral earth pressures, subsidence and foundation rotation and translation; gravity loads from walls, floors, and roofs; and vibrational loads from vehicular traffic, machinery and blasting.

Structural distress that results from these loads is usually manifested by fractures in the masonry units themselves, in the masonry joints, or at structural interfaces with other materials. This distress tends to accelerate deterioration, compromise the safety of the structure, and detract from its aesthetic qualities. The committee concluded that structural, diagnostic and remedial techniques for stone structures are in a relatively crude state. Existing professional information is not well disseminated, and there is little significant structural research taking place.

The dominant cause of deterioration of stone and masonry structures is water and moisture. Diagnosis of the presence of moisture can be carried out qualitatively with a simple portable instrument that measures electrical conductivity at the stone or masonry surface. A more quantitative portable instrument is the neutron moisture probe used in soil science and civil engineering. It is commercially available and could be used for nondestructive diagnostic surveys of structures. An infrared scanner has been used successfully to identify moisture-laden areas of large flat roofs (the moisture provided greater heat conduction, which is recognized by the infrared scan). The scanner costs about $10,000, but the savings effected by the diagnosis can make it cost-effective, the committee suggested. In addition technology used in the analysis and treatment of concrete highways is readily available for use in stone structures. A simple instrument used to measure vibration in roads could be used without modification to measure vibration in and movement in buildings.

Both the oil industry and NASA have techniques to measure density, porosity and water saturation of rocks, to make surface elemental analysis, and possibly to identify internal stresses and fractures. These techniques could be applied to dimension stone, but are currently far too complex and expensive for general use.

Oil industry research laboratories are seeking to extract oil from rock by reducing surface tension. If these researches prove successful some of the resulting technology may be useful in reducing the surface tension in the pores of rock and other masonry materials to discourage capillary action and rising damp.

The committee states that methods of cleaning stone structures requires more study. There are a number of successful treatments, they note, but a rational, systematic approach is lacking. Criteria for coatings and vapor barriers—their permeability, mechanical behavior, and compatibility with the substrate—require considerable research.

The question of consolidation of stone has implications beyond simply lining pores and hindering the migration of various constituents in the overall system. The questions that must be answered are, after an initial treatment is applied, should it be used again, or is the once-treated system now a completely different system requiring other treatments. Questions of compatibility of systems are not well understood.

Treatments are not permanent. Those that apply them must be prepared to take continuing responsibility for the stone's condition. Those responsible for initial treatments are often not responsible for subsequent maintenance, therefore, maintenance personnel must be made aware of the principles of preservation.

Vegetable, Animal, and Mineral Problems

J. Walter Roth, Director, Historic Preservation Staff, Public Building Service, General Services Administration notes that the most frequently encountered problem is the discoloration of stone and masonry. It is caused by the transfer of matter from adjacent substances as the result of runoff from dissimilar materials and dirt. The green stains on the marble bases of bronze statues or on the façades of buildings covered by copper roofs are apparent even to casual observers. Staining from unseen sources, such as flashing, cramps, conduits, gutters, reinforcement, wiring, and similar devices may puzzle the casual observer and alarm the concerned professional. To the former it is at most an instance of an unaccountable effect. To the latter it is an indication of imminent or eventual possible danger.

Vegetable impacts on masonry structures are familiar and frequent. They range from microcosmic mosses to clinging vines to unplanned hanging gardens growing out of crevices in buildings and monuments. There are also transitory problems, such as seasonal deposits of debris and damage from the rising roots or falling limbs of trees.

The impact of birds on masonry structures is probably the most varied and the most damaging. They roost on the newer, more streamlined buildings almost as comfortably as they do on the older, more highly ornamental ones. Their ability to adopt any kind of structure as a toilet facility seems unlimited. The residues from their acid-laden droppings, their nests, and even their corpses pose problems in the operation and maintenance of buildings and monuments.

Some Notes on the Physical Properties of Building Stone

Porosity and permeability are probably the most important physical properties in the study of rocks to understand the decay and corrosion of building and monumental stones according to geophysicist, Eugene C. Robertson of the U.S. Geological Survey. These properties describe the accessibility of water to the interior of the stones. Water in its three phases, vapor, liquid and ice, is perhaps the most important substance causing weathering and deterioration of stones. Thermal and mechanical properties are next in importance in the causes of decay of rocks. Their electrical and magnetic properties seem to have very little effect upon them.

Professor Erhard M. Winkler, geologist of Notre Dame University, states that stone decay is determined by the type of stone and the amount and source of moisture. The carbonate rocks—limestones, dolomites, and marbles—are attacked by moisture from the surface downward. Crystalline marble dissolves around the grains, resulting in sanding and a rough surface relief. Secondary layers and crusts of gypsum may form by dissolution and redisposition in the presence of sulfate, a process often aided by bacterial action. The decay of silicate minerals and rocks is very slow, except for tremolite in some dolomite marbles and black mica in granites and some marbles. Black mica may form brown blotches around mica flakes, whereas tremolite decays to soft talc leaving craterlike holes in marble. Granite rocks tend to separate into thin, even sheets parallel to the surface near ground level. Ground moisture combined with the action of salts and stress from the weight of the building causes this common spall, while the mineral components themselves remain unweathered.

The weathering and weathering rates of stone depend on the routes of travel and the amount of moisture. Corrosive rain and drizzle on the stone's surface; rising ground moisture of variable corrosiveness which is a vehicle for salt transport leading to efflorescence, subflorescence, and honeycombs; leaking indoor plumbing and gutters leading to uneven cleaning of the stone's surface and secondary deposits of calcite or gypsum, or both, are all causes and effects. And to this impressive list can be added the outward seepage of condensation water which leads to flaking, surface hardening and honeycombs in the rock.

Preventing the access of moisture is the most natural but most difficult way of preserving stone.

Mechanism of Masonry Decay Through Crystallization

The research of professor Seymour Z. Lewin of New York University indicates that one of the most common and extensive sources of deterioration of stone, brick, mortar, plaster, and concrete is the result of the crystallization phenomena that take place in pores, channels, and cracks in the material near its exposed surface. Liquid water deposits dissolved matter when it evaporates. This crystallization takes place when the dynamic balance between the rate of escape of water from the surface and the rate of resupply of water to the site of evaporation is upset. The former is a function of temperature, air humidity, and local air currents. The latter is controlled by surface tension, pore radii, viscosity, and the path length from the source of the solution to the site of evaporation.

The nature of this balance determines the form that decay will take. If the rate of resupply of solution to the surface is sufficient to keep pace with the rate of evaporation, the solute is deposited on the external surface of the masonry and is characterized as an efflorescence. If the rate of migration of solution through the pores of the masonry does not bring fresh liquid to the surface as rapidly as the vapor departs, a dry zone develops just beneath the surface. Solute is then deposited within the stone at the boundary between the wet and dry regions, generating spalls, flakes or blisters.

Professor Lewin has confirmed the validity of these insights in a variety of laboratory experiments and on site experiments.

REFERENCE

Sources as noted from *Conservation of Historic Stone Buildings and Monuments,* Report of the Committee on Conservation of Historic Stone Buildings and Monuments; National Materials Advisory Board, Commission on Engineering and Technical Systems National Research Council; National Academy Press, Washington, D.C., 1982

Diagnosis

One of the most important papers for those involved in building diagnosis was that of Baird M. Smith, Historical Architect, Heritage Conservation and Recreation Service of the U.S. Department of the Interior. Mr. Smith explored stone and brick deterioration and defects in building design as they related to building deterioration. His focus was on building practices used from the late 19th Century to the present time. Some of his observations follow.

Mr. Smith notes that there was tremendous competition among builders, architects and product manufacturers from 1880 to 1940. Many systems and products introduced at this time were proprietary with an attempt to keep methods and materials secret. Many products were also used without proper testing and installed with poor workmanship. Most systems and products were developed through trial and error methods. It was an age of exploitation of building materials and systems, Smith says, and we reap this doubtful legacy today in building failure.

It has been traditionally recognized as good building practice that when choosing a building stone or brick the weathering performance of the material should be matched to its expected exposure. Stone at the top of a wall or at the cornice must be more durable and capable of withstanding more severe weathering conditions than stone on the lower part of the wall.

In a competitive building market this practice has often been ignored. Stones were not placed where or how they were most needed. As proof of this today, deterioration of weak stones and bricks is common. The materials were simply not placed correctly.

Mr. Smith points out the contradiction in the use of ornamental carved stone which may not be practical if it is exposed to extreme weather. This is a logical incongruity. Soft stone is easiest to carve but for the reasons it is workable it is generally the least durable. A heavily carved balustrade at the top of a wall is the worst possible location for a soft stone. This contradiction has often been ignored in the past.

There are many chemical and physical incompatibilities between building materials. Some of these are well known, such as acid reaction on limestone or salts within masonry. Some less obvious but equally devastating incongruities are:

- Portland limestone and brick cause problems. Rainwater washing down limestone onto brick will stain the surface and damage the brick. Little can be done to arrest this process.
- Ferrous oxide in portalnd cement will always stain limestone, marble, and some sandstones. Nonstaining portland cement has been developed to eliminate this problem.
- Oxidation of iron, steel and copper can cause both staining and physical damage to stone and brick. Careful attention must be paid to isolating these materials in contact with masonry.

Overall building profiles have changed over the years. A comparison of an 18th Century wall section of a stone building with one from the mid-20th Century shows an evolution away from the conscious control of rainwater on the face of the building. The early building would most probably have a projecting cornice, pediments over each window, projecting belt courses and a water table. All these architectural features were intended to direct rainwater off the building façade, minimizing staining, weathering and other moisture problems. Façades of recent buildings, in contrast, are comparatively flat. There is little or no attention paid to the control of rainwater. As a result moisture problems are quite common.

Walls

The basic construction of the building walls have changed. They have evolved from simple bearing to "cavity" to nonbearing curtain walls. In most instances, nonbearing walls replace bearing walls except in residential scale buildings beginning in the 1880s and are found in common use by the turn of the century.

Bearing Walls

Stone masonry bearing walls involved the laying of cut stones of from 10 inches to over 3 feet in width on top of each other with bedding planes in a natural horizontal position as they were quarried.

Brick masonry bearing walls were built of two, three, or four wythes solidly packed with mortar. Lime mortar was normally soft. Its purpose was to form a uniform bearing between masonry units. It in effect held the units apart. It did not hold them together, for lime mortar was not strongly adhesive.

Individual projecting stones were connected with dowels, anchors, or cramps where they were set to keep the water off the wall and away from openings at cornices or over windows. Before the widespread availability of steel in the late nineteenth century metal anchors were generally made of wrought iron. They were set into the stone with molten lead, a sulfur and sand mix or a mortar grout. Wrought iron generally resisted severe corrosion as long as it was buried sufficiently within the wall to remain dry. Severe corrosion did occur when the anchors were exposed to the weather that jeopardized the integrity of the connection. The result could be failure or collapse of the stonework.

It is now recognized that wrought iron, cast iron, or steel anchors set in sulfur or in mortar can severely corrode. If moisture reacts with the sulfur forming sulfuric acid or with the mortar forming carbonic acid corrosion and damage is accelerated.

Cavity Bearing Walls

It was widely recognized by the early nineteenth century that walls of solid masonry were susceptible to moisture penetration from hard driving rain. To prevent through penetration a vertical air space or cavity of from two to four inches was introduced into the wall. The wall had two sections, an outer one of dressed stone or hard-fired bricks and an inner wall of rubble stone or irregular brick. The two thicknesses were laid independently yet simultaneously. Generally both bore the weight of the wall above. An exception was very thick walls of more than 24 inches, where weight was carried on the inner wall.

The outer wall was designed to take the brunt of wind rain or airborne objects. The cavity provided an outlet for moisture that penetrated the outer wall. The moisture flowed to the base of the cavity and was expelled through weep holes or other small openings to the exterior. Wall sections were tied together with various types of metal ties or bond stones, bricks or structural terra-cotta turned at right angles to the face of the wall. These bridged the cavity. Metal ties were first made of wrought and cast iron, then by the mid-19th Century, galvanized iron, painted iron, iron dipped in tar, and even copper or bronze was used. Early literature shows that some of these cavity-wall ties were shaped to eliminate flat surfaces which were potential catch basins for moisture in the cavity.

There are a number of problems typical of cavity walls, Smith points out. Differential settlement between inner and outer wall sections can take place particularly if the loads of the building are framed into the inner wall section. This will weaken or break the ties between the two sections thus jeopardizing the wall's structural integrity. The most common problem, recognized from the very beginning, is that of keeping the cavity clean. It tends to become clogged during construction with mortar and other debris. Mortar will also plug weep holes or rest on top of ties creating catch basins for moisture and thermal links between inside and outer walls. Metal ties, whether painted or galvanized, are very susceptible to corrosion because the air cavity is very moist.

During the first part of the 20th Century skeleton steel and concrete frame construction brought the curtain wall. These walls, Smith notes, were rarely thicker than 12 inches. Their weight was carried on each floor or on every other floor by the floor framing. This was a time in which a host of new technologies were introduced into building and new kinds of building materials used. Elevators, skyscrapers, fireproof construction and lightweight building skins were some of the major elements of this change.

The introduction of new materials proves of great significance to us today in our efforts to repair or save some of these buildings. Bricks and stone are quite heavy and capable of supporting themselves. They were not the best material for curtain walls if the key function of the wall was to provide a lightweight enclosure. Compressive strength was no longer an issue. Sandlime bricks, architectural terra-cotta, cast stone, glass, metal, and sandwich panels were introduced.

Because these new materials were lightweight and non–load bearing they had to be extensively fastened to masonry backing and to structural supports. The types of anchors employed varied greatly, both in shape and in material. But all were intended to be noncorrosive. The most common materials used at this time were painted and galvanized iron and steel ties and wires. In some instances bronze and copper ties and wires were used and more recently various plastic materials have been employed.

In the first years of this century, painted steel was considered noncorrosive; this proved, perhaps disastrously, to be untrue. Galvanized steel was substituted as the minimum requirement. After World War II stainless steel was introduced but galvanized and painted iron were never discontinued. They are used today but it is now accepted that they do not have the life potential of stainless steel. Some of the early stainless steels also proved to be corrosive. It is therefore essential today to be sure that the anchoring device specified is capable of withstanding the corrosive conditions within the wall for the life of the structure.

Another problem is that facing stone on curtain walls is very thin. It can be two inches or less in thickness. It is generally face bedded rather than bedded in its natural plane. The result is delamination and extensive surface spalling.

A second bothersome issue is the wall cavity. There is no question that cavity walls are efficient in preventing moisture penetration into the building. In curtain-wall construction the cavity was treated as it had been in bearing walls, consisting usually of 4 inches outer wall veneer of brick, stone or terra-cotta then an air cavity of from 2 to 4 inches and then the inner masonry backup for a total wall thickness of about 12 inches. This was the usual cavity wall construction but new details had to be devised to sheath structural iron with masonry. Structural iron was encased with stone. The resulting cavity was vented to the outside with open joints or weep holes and the stone anchored with metal ties. Although the encased iron and steel was painted corrosion was still common. To correct this problem, the cavity was eliminated and all steel or iron was thoroughly covered or encased with mortar grout. However, the resulting solid masonry wall permitted moisture penetration that disrupted interior wall finishes. To solve this problem a cavity was introduced just inside the finished wall surface. Plaster lath was nailed to furring strips resulting in a ¾-inch cavity between the exterior wall and interior finish plaster.

Another change in cavity wall construction was the addition of flashing within the wall cavity at selected points to control runoff and protect critical connections and anchors.

Another critical problem is that mortar between masonry units moves. Lime mortar expands and contracts with changes of temperature or moisture content. In a bearing wall the mortar joint is in compression and expansion or contraction of materials has little effect on the structural integrity of the wall. With the evolution of the curtain wall, builders changed to stronger, waterproof, portland cement mortar.

The reason for this change, Smith says, is not quite clear. There was a reference to lime mortar's failing in fire and a need to make the mortar waterproof, possibly to protect the anchoring devices and structural framework.

However it is recognized today that high-strength cement mortars fail to prevent rain penetration. This may be due to the following reasons:

- Portland cement mortars shrink upon setting. A curtain wall is not always in compression. Minute horizontal fractures occur from shrinkage. Often an elastomeric joint sealant or caulking is used in selected horizontal joints around a building, generally at the top of each section of the curtain wall where it joins a floor or spandrel beam. It is also extremely difficult to adequately mortar the last brick course of the wall in this position. It has been found that through the years these joints open creating an entry point for rain.
- In tall buildings, lateral wind loads place tensile forces on the curtain wall, opening joints on the windward side of the building. This breaks the mortar bond, creating thousands of entry points for rain.
- Portland cement reacts with airborne sulfates, resulting in "sulfate attack," causing expansion of mortar and disruption of mortar bond.

These problems aggregated create the potential for severe deterioration of curtain wall materials. especially iron and steel anchorage and structural systems.

To counter these potential problems, modern high-rise buildings rely heavily on flexible joint sealants. If the best materials or techniques are not used and the joints are not maintained, the results can be disastrous.

Parapets

A parapet is susceptible to deterioration and damage. It is is free standing and not warmed by interior building heat. It is exposed to the worst weather conditions, including harsh freeze-thaw cycling.

Builders have usually recognized the vulnerability of parapets and go to great lengths to protect parapet walls. These walls are topped with capstones which are provided with sloping surfaces to create a water wash. The capstone extends beyond the wall and is provided with a reglet or drip on the underside. Joints in capstones are vulnerable since they face up. They are sometimes packed with lead or tar, or covered with galvanized iron or copper.

These practices have not changed appreciably from medieval times. Yet the need to provide for horizontal expansion due to thermal loads has not always been considered. As a result capstones can become dislodged, creating open vertical joints for the entrance of water into the wall.

Cornices and Overhangs

In simple bearing wall construction the cornice is merely a corbelled or cantilevered stone. When buildings become higher the desire to preserve proportional relationships meant that cornices became massive overhangs of stone or terracotta suspended from the structural frame of the building. These metal fastenings have not always been designed to protect them adequately from corrosion nor have noncorrosive materials been used.

Roof Drainage

A change in roof drainage patterns has taken place. The most common early technique was to pitch the roof, hip, gable, or gambrel which directed rainwater off and away from the building. Later gutters and downspouts were introduced to control water and transport it down and away from the base of the wall.

Eventually as buildings became taller, downspouts, of cast iron were built into the wall. This created several problems. If the downspout was too close to the outside surface of the wall and became clogged it froze, fracturing the cast iron and damaging the wall. If clogged it was extremely difficult to clean and the repair of leakages is a problem.

Gutters are also a constant source of masonry problems. If hidden they are usually lined with copper or lead but these metals corrode and crack. They are corroded by acid from vegetation such as leaves that accumulate. As a result the joints open, and water enters the cornice and top of the wall, causing damage.

Sills

A common location for masonry deterioration and damage is at window and door openings. These openings penetrate wall enclosures. Water may enter the center portions of the wall through them. Generally damage occurs at heads and sills. Sills must be sloped outward with a drip or reglet. The sill should extend into the masonry on either side of the opening and project beyond it to drip water away or staining and deterioration will result. Because sills are exposed they are susceptible to deterioration especially from freeze/thaw cycles. If face bedded stone is used serious delamination and spalling result.

Lintels

The head or lintel above a window or door opening is a common problem area. Traditionally lintels were of wood or masonry. They were either a single stone or piece of wood spanning the full width of the opening. It could also be a combination of stones or bricks forming a flat, segmental or pointed arch. But no matter what the shape it carried the weight of the masonry wall above. With the availability of iron and steel angles, masonry arches were eliminated. Because most of these angles are partially exposed to the weather the potential for corrosion is great. In most cases there is little, or no, provision made to prevent rainwater from coming down the face of the wall above and under the lintel. This water flows back to the angle increasing its corrosion potential.

Base of Walls

Problems associated with groundwater, surface water, and related salt or freeze/thaw problems have been reasonably well understood since the mid-19th Century. Building practices designed to avoid such problems have remained largely unchanged from that time to this.

Problems in older masonry foundations generally relate to a breakdown in materials or systems intended to keep water out of basements. Outside surfaces of foundation walls were often coated with cement pargeting or asphaltic mastics. These coatings, over time, can become dislodged and no longer block moisture penetration. The result is damage to the foundation, especially to mortar joints.

Footing and foundation drains are quite common during the 20th Century. These drains of clay tile or masonry often become clogged with silt, tree roots or organic or animal matter which reduces or eliminates drainage of groundwater. The result is damage to the foundation, perhaps posing a jeopardy to the building's stability.

REFERENCE

Diagnosis of Nonstructural Problems in Historic Masonry Buildings, pp. 211–232 in *Conservation of Historic Stone Buildings and Monuments,* by Baird M. Smith, Historical Architect, Heritage Conservation and Recreation Service, U.S. Department of the Interior. Published by National Academy Press, Washington, D.C., 1982

STONE CLASSIFICATIONS

Almost every kind of rock can be used as dimension stone. The principal determinants of usage are aesthetic appeal, physical properties, and resistance to weathering. The geological definition of a rock is based on its chemistry, fabric, and mineralogy. These attributes are also the principal determinants of the stone's properties. The American Society for Testing and Materials (ASTM) has adopted standard definitions for the most commonly used commercial building dimension stones. The following listing of building stones by Norman Herz, professor of Geology, University of Georgia, describes them both from geological and ASTM points of view.

Rocks are divided into three overlapping genetic groups:

- Sedimentary rocks, such as limestone and sandstone.
- Igneous rocks, such as granite and diabase (traprock).
- Metamorphic rocks, such as marble and slate.

Limestone

ASTM defines limestone as a rock of sedimentary origin, composed principally of calcium carbonate (calcite) or the double carbonate of calcium and magnesium (dolomite). The textures vary greatly, from uniform grain size and color to a cemented-shell mash. Oolitic limestone, a popular building stone in this country, Britain, and France, consists of cemented rounded grains of calcite or aragonite generally under 2 mm in diameter. Some limestones have varying amounts of other material, such as quartz sand or clay mixed in with the carbonate minerals. Most limestones are formed of shells or reworked shell fragments, although many commercial limestones, including oolitic and very fine-grained compact varieties, are chemical precipitates.

Sandstone

ASTM defines sandstone as a "consolidated sand in which the grains are composed chiefly of quartz and feldspar, of fragmental texture, and with various interstitial cementing materials, including silica, iron oxides, calcite, or clay." Commercially used sandstone is a clastic sediment consisting almost entirely of quartz grains, $1/16$ to 2 mm in diameter, with various types of cementing material. Enough voids generally remain in the rock to give it considerable permeability and porosity. In the United States, commercially available sandstones include the well-known brownstone, an arkosic sandstone that is rich in feldspar grains and was quarried in the Triassic basins of the eastern states.

Travertine

Travertine is a variety of limestone deposited from solution in groundwaters and surface waters. When it occurs hard and compact and in extensive beds as around

Rome, Italy, it can be quarried and used as an attractive building stone. It is generally variegated gray and white or buff, with irregularly shaped pores distributed throughout the ground-mass.

Granite

Commercial granite includes almost all rocks of igneous origin. True granites consist of alkali feldspars and quartz with varying amounts of other minerals, such as micas and hornblende, in an interlocking and granular texture, and with all mineral constituents visible to the naked eye. Geologically granite is distinguished from other rocks that it resembles, such as granodiorite, quartz monozonite, and syenite, on the basis of the percentages of quartz, potassium feldspar, and plagioclase feldspar. This distinction is not made commercially; in fact, black fine-grained igneous rocks, such as basalt or diabase, are commonly called "black granite." Other dark "granites" include rocks that, petrographically, are anorthosite, gabbro, syenite, and charnockite.

Marble

According to ASTM, commercial marble includes all crystalline rocks composed predominantly of calcite, dolomite, or serpentine and capable of taking a high polish. Geologically, marble is considered only as a metamorphic rock formed by the recrystallization of limestone or dolomite under relatively high heat and pressure. Thus, in addition to geological marble, commercial marble includes many crystalline limestones, travertine, and serpentine, a metamorphosed ultramafic rock. In the metamorphic process, original sedimentary features, except for bedding, which is preserved as compositional layering, are destroyed. The original minerals, calcite and dolomite, are recrystalized in an interlocking mosaic texture, and the impurities form magnesium and iron silicates. The color of many marbles is due to these accessory minerals, such as talc, chlorite, amphiboles, and pyroxenes, as well as iron oxides, hydroxides, sulfides and graphite.

Slate

ASTM requires a slate to possess an excellent parallel clevage that allows the rock to be split with relative ease into thin slabs. Slate is a metamorphosed rock derived from argillaceous sediments consisting of extremely fine-grained quartz, the dominant mineral, and mica and other platy minerals. The color of slate is generally determined by the oxidation state of the iron or the presence of graphite or pyrite.

Other Types of Stone

A great variety of other types of rocks are sold commercially, including: (a) quartzite—a metamorphosed sandstone consisting almost entirely of quartz and utilized locally, as the Sioux Falls quartzite of South Dakota and the Baraboo quartzite of Wisconsin; (b) greenstone—defined by ASTM as a metamorphic rock principally containing chlorite, epidote, or actinolite; (c) basalt or traprock—a microcrystalline volcanic or dike rock that consists primarily of pyroxene and calcic plagioclase (the stark black churches of the Auvergne of Central France are largely made of basalt); and (d) obsidian—a volcanic glass that commercially includes pumice in the United States, has low density because of its frothy texture, and can be easily shaped with hand tools.

REFERENCES

Conservation of Historic Stone Buildings and Monuments: Report of the Committee on Conservation of Historic Stone Buildings and Monuments, National Materials Advisory Board, Commission on Engineering and Technical Systems; National Research Council, National Academy Press, Washington, D.C., 1982.

Feilden, Bernard M.: "The Principles of Conservation" in *Conservation of Historic Stone Buildings and Monuments,* National Academy Press, Washington, D.C., 1982.

Roth, J. Walter: "Some Illustrative Preservation Problems and Treatments in Washington, D.C." in *Conservation of Historic Stone Buildings and Monuments,* National Academy Press, Washington, D. C., 1982.

Herz, Norman: "Geological Sources of Building Stone" in *Conservation of Historic Stone Buildings and Monuments,* National Academy Press, Washington, D.C., 1982.

Robertson, Eugene C.: "Physical Properties of Building Stone" in *Conservation of Historic Stone Buildings and Monuments,* National Academy Press, Washington, D.C., 1982.

Walsh, Joseph B.: "Deformation and Fracture of Rock" in *Conservation of Historic Stone Buildings and Monuments,* National Academy Press, Washington, D.C., 1982.

Winkler, Erhard M.: "Problems in the Deterioration of Stone" in *Conservation of Historic Stone Buildings and Monuments,* National Academy Press, Washington, D.C., 1982.

Lewin, Seymour Z.: "The Mechanism of Masonry Decay Through Crystallization" in *Conservation of Historic Stone Buildings and Monuments,* National Academy Press, Washington, D.C., 1982.

Robinson, Gilbert C.: "Characterization of Bricks and Their Resistance to Deterioration Mechanisms" in *Conservation of Historic Stone Buildings and Monuments,* National Academy Press, Washington, D.C., 1982.

Smith, Baird M.: "Diagnosis of Nonstructural Problems in Historic Masonry Buildings" in *Conservation of Historic Stone Buildings and Monuments,* National Academy Press, Washington, D.C., 1982.

Fitzsimons, Neal and Colville, James: "Diagnosis and Prognosis of Structural Integrity" in *Conservation of Historic Stone Buildings and Monuments,* National Academy Press, Washington, D.C., 1982.

Stone Decay—Checklist

Stone deterioration is closely related to other parts and other building materials. The conditions surrounding the building, brick backup walls, mortar, iron clamps, gutters, cornices, foundation drainage, and salts leaching through the soil can all be contributors.

Stone deterioration begins with an exertion of some form of unusual pressure exerted by either physical or chemical forces. Water penetration is almost always the source or the vehicle of the problem. Moderate exposure to rainwater is inevitable and to some extent beneficial. Stone damage occurs with cycles of heavy saturation and drying, often caused by leaking gutters, defective cornices, blocked drains, or clinging vegetation. The most common causes of stone decay are:

Stonehenge. (Photo Courtesy of CBS News)

Gravestone, Trinity Church Graveyard, New York City. (Photo by F. Wilson)

Shaling cliffside, Millbrook, New York. (Photo by F. Wilson)

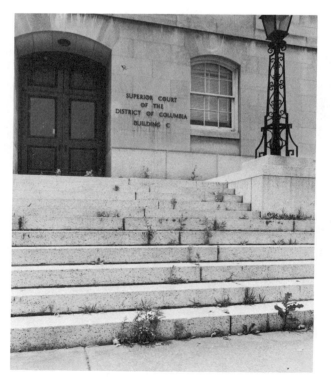

Stone and plants, Court Building, Washington, D.C. (Photo by
F. Wilson)

1. The freeze-thaw cycle. Water drawn deep into the stone by capillary action causes mechanical stress on the pore walls during freezing, a major cause of stone deterioration in harsh climates. Freezing and thawing affects limestone and sandstone more severely than relatively impervious stones like granite and marble.

2. Salt crystallization. Soluble salts are carried with water into the pore network where they form crystals. Salts increase in volume as they crystallize. Their expansion damages the internal structure of the stone, and salt deposits efflorescence on the surface of the stone. Salts may be absorbed from the soil if plinths and water tables are not constructed of an impervious stone. More frequently, salts originate in mortars containing Portland cement or originate in brick or the backing material to which a stone facing is attached. Efflorescence frequently appears on new stone facing placed against a damp backing material. Salts from bird droppings deposited on cornices and sills and sodium chloride scattered to melt on steps will also aggravate stone decay.

3. Acids deposited on calcareous stones, like limestone, marble, and some sandstones, gradually dissolve the stone. The result is roughening of the surface of the stone, separation of the bedding planes where soft seams were washed away, erosion around harder fossil fragments, and loss of detail in carvings. On

Deteriorating stone pediment and stone wall, City Hall office.

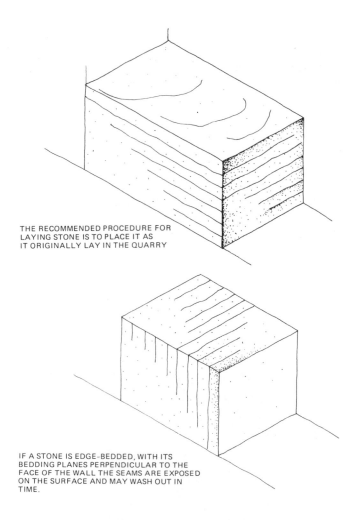

THE RECOMMENDED PROCEDURE FOR
LAYING STONE IS TO PLACE IT AS
IT ORIGINALLY LAY IN THE QUARRY

IF A STONE IS EDGE-BEDDED, WITH ITS
BEDDING PLANES PERPENDICULAR TO THE
FACE OF THE WALL THE SEAMS ARE EXPOSED
ON THE SURFACE AND MAY WASH OUT IN
TIME.

sheltered surfaces not washed by the rain, hard impermeable skins of calcium sulfate and sulfur dioxide from the atmosphere form on stone and can cause exfoliation, blistering, and spalling due to differences in moisture and thermal movements between the skin and the stone.

4. Moss, lichens, vines, and thick shrubbery around a stone foundation tend to harbor moisture in stone by trapping it and preventing evaporation. Some plants also secrete acids which have a mild solvent effect, particularly on limestone and marble. But probably the greatest damage is caused by roots, which gradually open joints and dislodge particles of mortar as they expand.

5. In cases of severe structural settlement, sound stones will crack at points of weakness, such as lintels and sills of doors and windows.

6. Face bedding. The durability of sedimentary stones, such as limestone and sandstone can be drastically affected by the incorrect placement of the blocks in the building. Stone blocks should be laid in the same position as the stone originally lay in the quarry. On its natural bed the stone's bedding planes are horizontal. If face bedded, or laid with the bedding planes vertical and parallel

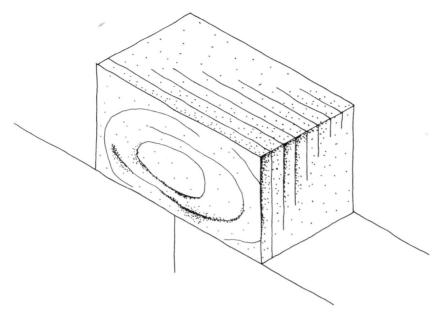

If the stone is face-bedded, with its bedding planes placed parallel to the face of the wall it tends to scale off in layers. (Drawings after Technical Series No. 5; Preservation League of New York State) Building, New York City. (Photo by F. Wilson)

to the face of the wall, the surface of the stone will scale in sheets or layers. Face bedding was quite common in 19th-Century construction; stone for columns or door jambs was often placed on end to take advantage of its greater length.

7. Edge bedded stones are laid with the bedding planes vertical but perpendicular to the face of the wall. In time the seams of the exposed surface of the stone may wash out between the laminations. Edge bedding is acceptable in cornices and string courses which would erode rapidly if the stone were laid in its natural bed.

8. Spaces between stones that are inadequately sealed allow water to penetrate deep into the masonry. Open joints are caused by settlement and by the mortar's inability to withstand the physical pressure and expansion of stone. Joint problems arise due to inadequate mortar in the joint; eroded mortar lacked enough lime to make it plastic.

9. Splitting caused by rusting of embedded ironwork is a frequent problem with railings set into stone steps as well as with concealed metal dowels and clamps inserted during construction, particularly in copings. As iron corrodes, its volume increases, it exerts pressure on the hole in which the iron member was fitted, causing damage to the stone.

REFERENCE

Gilder, Cornelia Brooke, Technical Series No. 5, Preservation League of N.Y. State.

V. Wood

Dielectric-Type Moisture Meter

Resistance-Type Moisture Meter

Electrical Resistance Probe

Ultrasonic and Impact-Induced Longitudinal-Wave Pulse Velocity Equipment

Stress-Wave Propagation Equipment

Oven-Dry Test

Radiography

Decay Resistance of Wood

Tables

REVIEW WOOD CHARACTERISTICS

Certain of the growth characteristics of wood affect its grading and use. These include knots, shakes, pitch pockets, and splits.

Knots are one of the most common growth defects formed when branches of a tree, which originate in the center of the pith, become enclosed within the wood during growth. If a branch dies and later falls off, the dead stubs are incorporated into the wood, become overgrown with new wood, and form knots when cut into lumber.

Shakes are grain separation occurring between annual growth rings. They extend along the grain parallel to the height of the tree. Shakes seldom develop in lumber unless present in the tree before it is felled.

Pitch pockets are small, well-defined grain separations that may contain solid or liquid resin. Pitch pockets may be found in Douglas fir, western larch, pine, and spruce.

The properties of wood are determined by its physical and chemical composition. The most characteristic feature of wood is its cellular structure. Wood cells or fibers are primarily cellulose, cemented together with a material called lignin. The wood structure is approximately 70% cellulose, from 12 to 28% lignin and up to 1% ash-forming materials. These constituents are responsible for wood's hygroscopic properties, its susceptibility to decay, and its strength. The bond between individual fibers is so strong that when tested in tension they commonly tear apart rather than separate. The remaining wood characteristics are extractives that give different species distinctive characteristics such as color, odor, and natural resistance to decay.

Hygroscopic Properties

Wood is hygroscopic, it expands when it absorbs moisture and shrinks when it dries or loses moisture. This property affects practically every single end use of wood. Although the wet (green) condition is normal for wood throughout its life as a tree, most products made of wood require that it be used dry; therefore, seasoning by drying to an acceptable moisture content is necessary.

The moisture content of wood is the weight of the water it contains expressed as a percentage of the weight of the wood when oven-dry. The oven drying determines moisture content (m.c.) in the laboratory. On the building site the moisture meter gives an instantaneous reading by measuring the resistance to current flow between two pins driven into the wood.

Pin Knot — less than ½ in.
Small Knot — less than 1½ in.
Large Knot — more than 1½ in.

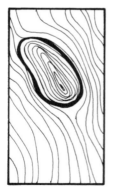

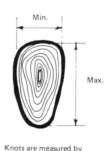

Min.

Max.

Knots are measured by
averaging the maximum
and minimum diameters.

ROUND KNOT
(Sawn perpendicular
to the branch)

OVAL KNOT
(Sawn diagonally
to the branch)

SPIKE KNOT
(Sawn parallel
to the branch)

Knots may be considered sound if free from decay, or unsound if decay has occurred; tight if it will firmly retain its place, or loose if it cannot be relied upon to remain in place.

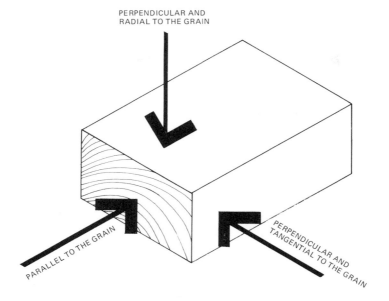

PERPENDICULAR AND
RADIAL TO THE GRAIN

PARALLEL TO THE GRAIN

PERPENDICULAR AND
TANGENTIAL TO THE GRAIN

Weathering wood siding, miners cabin Northern New Mexico (Photo by F. Wilson)

Fiber saturation in green wood is present in two forms; in the cell cavities as free water and within the cell fibers as absorbed water. As the wood dries, the cell fibers do not give off their absorbed water until all the free water is gone and adjacent cell cavities are empty. The point at which the fibers are still fully saturated, but the cell cavities are empty, is called the fiber saturation point. This occurs in most species at about 30% m.c. This condition represents the point at which shrinkage begins. Even though green lumber may be cut with moisture contents as high as 200% with subsequent drying to the fiber saturation point around 30% m.c. no shrinkage of the wood occurs until after the cell cavities lose their free water. When the cell fibers give off their absorbed water and begin to constrict the wood shrinks.

The total shrinkage of wood takes place between the fiber saturation point and a theoretical moisture content of 0%. Within this range shrinkage is proportional to moisture loss. For every 1% loss or gain in moisture content below 30%, the wood shrinks or swells, respectively, about one-thirtieth of its total shrinkage potential. At 15% m.c. the wood has experienced half its total shrinkage. Wood in service almost never reaches a moisture content of 0% because of the presence of water vapor in the surrounding atmosphere. Therefore the total shrinkage possible is far less significant than probable actual shrinkage.

Equilibrium Moisture Content

After a tree is felled and while it is being cut into lumber its moisture content drops as the moisture in the wood is lost to the surrounding air. Wood moisture eventually comes to a point of equilibrium with air moisture. At this point the moisture content of the wood is called the *equilibrium moisture content*. There is a constant relation between atmospheric temperature and humidity to the equilibrium moisture content of wood.

The practical significance of knowing the equilibrium moisture content is that it allows prediction of the moisture content wood will attain in service. For wood to experience only minor dimensional changes it should, ideally, be fabricated and installed at a moisture content as close as possible to the equilibrium moisture content it will experience in service.

Structural Properties

Wood is fibrous. The length of fibers varies from about ½₅ in. in hardwoods to from ⅛–⅓ in. in softwoods. The strength of wood does not depend on the length of the fibers, but on the thickness of their walls and their direction relative to applied loads. Wood fiber length parallels the vertical axis of the tree. The strength of wood parallel to these fibers is quite different from its strength perpendicular to them.

Strength values of wood are determined by testing small, clear specimens in bending, compression, shear, etc., and assigning basic strength values to the different species. Test pieces provide theoretical values, lower working stresses termed, allowable unit stresses, are derived to consider the effect of knots, slope-of-grain, checks, or other strength-reducing characteristics.

(Photo by F. Wilson)

Materials like wood which expand and contract markedly with changes in moisture content, experience serious internal strains upon drying. Wood exhibits different shrinkages in the three principal directions of the tree, namely, along the length (or grain), radially, and tangentially with respect to the log cross section. The changes are about 0.1, 5, and 7%, respectively, from green to oven-dry. Even if a log is dried slowly, it will almost always crack, since circumferential shrinkage is greater than radial. Cracking is avoided by cutting the log into lumber, followed by slow and careful drying.

Two different materials having different properties placed together may load and strain each other with change of load, temperature, or moisture content. A common example is in the balancing of a sheet of material such as plywood with uneven piles. Two-ply material warps so badly as to be unusable. Cracking and checking of plywood veneers occurs if fabrication takes place at moisture contents above those at which the material will be used.

The daily, weekly and seasonal moisture changes in wood siding can lead to dimensional changes transverse to the grain in excess of 1%. Such changes occur in all log buildings, with some attendant difficulties when wood elements are applied to logs across the grain. Horizontal logs are self-adjusting to shrinkage, while cracks are bound to open up between logs placed vertically.

Problems will occur in cracking of wide weather boards, if they are solidly nailed at both edges. The Norwegians have special nailing methods for wide board siding to avoid this. The wide boards are nailed only in the center, while the battens are held by crossed nails passing between the gaps in the wide boards.

These and other problems occur with wood because it exhibits a relatively very high shrinkage and expansion across the grain with changes in moisture content. It would be very much less useful, however, if it exhibited similar movement along its length instead of the low values of 0.1% from wet to dry.

Recognition of the unique differential shrinkage characteristics of wood has been a prime influence in the development of traditional practices in wood construction. Light wood framing makes use of members in the more stable along-the-grain direction, with only minimum involvement of thin across-the-grain members such as plates and sills. Simple overlapping units such as shingles and siding allow for movements resulting from changing moisture conditions, without sacrificing function, by minimizing evidence of movement. Narrow board flooring distributes total shrinkage among a large number of joints so that it is least noticeable and surface irregularities from cupping and other warping effects are minimized.

The change in dimension resulting from changing moisture content for wood and even for plywood are much larger than for most other building materials, although some clay soils exhibit volumetric shrinkage of as much as 10% on drying from the saturated condition.

Allowable unit stresses for wood are based on normal duration of loading. Values for normal loading generally are used for designing members to carry dead loads such as floor or roof construction and live loads due to occupants, equipment and furnishings. For loads of brief duration such as those produced by snow, earthquake, or wind, higher working stresses are used.

EVALUATION OF STRUCTURAL TIMBER

The evaluation of wood structural members must identify the characteristics of the wood's strengths, such as density, knots, and moisture content that define its structural performance.

In contrast to man-formed materials such as concrete and metal which can be controlled in manufacture, timber is segregated into use classifications (grading) by growth characteristics based on its predicted performance in the structure. The building industry's control over the quality of wood is limited to the proper application of wood classifications to their appropriate structural use.

In an existing structure there are three questions which must be asked about wood performance:

1. Is the member properly specified according to grade and size?
2. Is the joining structurally appropriate?
3. Are there evidences of degradation of the wood due to moisture, temperature, chemical reaction, plant growth, or rot?

Grade markings stamped on the lumber at the mill are valuable aids in evaluating structural members. These can be related to a recommended design value by reference to the National Design Specification for Wood Construction or other relevant documents. They determine the quality and strength properties of the existing timber. The difficulty is finding them, which might require removing the finish of wood members. This is not customarily done because of the considerable cost of refinishing.

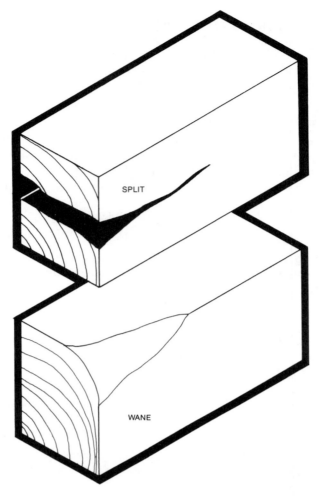

KNOTS-Are formed when branches of the tree originating in the center at the pitch become enclosed within the wood during subsequent growth. If a branch dies and later falls off, the dead stubs are incorporated in the wood. They become overgrown with new wood and form knots when cut into lumber.

SHAKES-A grain seperation occurring between annual growth rings, running along the grain parallel to the height of the tree. Shakes seldom develop in lumber unless present in the tree before it is felled.

PITCH POCKETS-A pitch pocket is a small, well-defined grain separation that may contain solid or liquid resin. Pitch pockets are found in Douglas Fir, Western Larch, Pine and Spruce.

CHECK-A separation of the wood across the growth rings.

SPLITS-Lengthwise grain separation extending through the piece from one surface to the other.

WANE-The presence of bark or absence of wood.

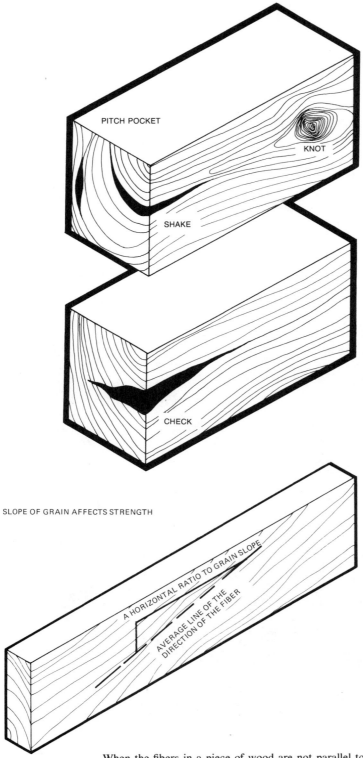

PITCH POCKET

KNOT

SHAKE

CHECK

SLOPE OF GRAIN AFFECTS STRENGTH

A HORIZONTAL RATIO TO GRAIN SLOPE

AVERAGE LINE OF THE DIRECTION OF THE FIBER

When the fibers in a piece of wood are not parallel to the edges the diagonal grain may have caused several problems. These could be warpage, shrinkage at the end, torn or lifted grain. The member is also weaker than a piece with parallel grain if it is loaded perpendicular to the grain direction.

If grade marks are not discernible it may be necessary to engage a wood evaluator experienced in identifying and grading wood products.

DECAY IN BUILDINGS

Serious decay problems in buildings are almost always a sign of faulty design or construction or lack of reasonable care in the handling of wood.

Construction principles that ensure long service and avoid decay include the following: (1) Build with dry lumber, free of incipient decay and not exceeding the amounts of mold and blue stain permitted by standard grading rules. (2) Use designs that will keep the wood dry and accelerate rain runoff. (3) For parts exposed to above-ground decay hazards, use wood treated with a preservative or heartwood of a decay-resistant species. (4) For the high-hazard situation associated with ground contact, use pressure-treated wood.

A building site that is dry or for which drainage is provided will reduce the possibility of decay. Stumps, wood debris, stakes, or wood concrete forms frequently lead to decay and termite attack if left under or near a building.

Damage due to water and termite infestation, Hawaiian Islands. (Photo by F. Wilson)

Unseasoned wood may be infected because of improper handling at the sawmill or retail yard, or after delivery on the job. Unseasoned or infected wood should not be enclosed until it is thoroughly dried.

Untreated wood parts of substructures should not be permitted to contact the soil. A minimum of 8 in. clearance between soil and framing and 6 in. between soil and siding is recommended. An exception may be made for certain temporary constructions. If contact with soil is necessary, the wood should be pressure treated.

Sill plates and other wood resting on a concrete-slab foundation generally should be pressure treated, and additionally protected by installing beneath the slab a moisture-resistant membrane of heavy asphalt roll roofing or polyethlene. Girder and joist openings in masonry walls should be big enough to assure an air space around the ends of these wood members; if the members are below the outside soil level, moisture proofing of the outer face of the wall is essential.

In the crawl space of basementless buildings on damp ground, wetting of wood by condensation during cold weather may result in serious decay damage. However, serious condensation leading to decay can be prevented by providing open-

Damage due to water and termite infestation, Hawaiian Islands. (Photo by F. Wilson)

Washington, D.C. (Photo by F. Wilson)

Plainfield, N.J. (Photo by F. Wilson)

ings on opposite sides of the foundation walls for cross ventilation or by laying heavy roll roofing or polyethlene on the soil; both provisions may be helpful in very wet situations. To facilitate inspection and ventilation of the crawl space, at least an 18-in. clearance should be left under wood joists.

Porches, exterior steps, and platforms present a decay hazard that cannot be fully avoided by construction practices. Therefore, in wetter climates the use of preservative treated wood or heartwood of a durable species usually is advisable for such items.

Protection from entrance or retention of rainwater or condensation in walls and roofs will prevent the development of decay in these areas. A fairly wide roof overhang (2 ft) with gutters or downspouts that are never permitted to clog is very desirable. Sheathing papers under the siding should be of a "breathing" or vapor-permeable type (asphalt paper not exceeding 15 lb weight). Vapor barriers should be near the warm face of the walls and ceilings. Roofs must be kept tight, and cross ventilation in attics is desirable. The use of sound, dry lumber is important in all parts of the buildings.

When service conditions in a building are such that the wood cannot be kept dry, as in textile mills, pulp and paper mills, and cold-storage plants, lumber properly treated with an approved preservative or lumber containing all heartwood of a naturally decay-resistant species should be used.

In making repairs necessitated by decay, every effort should be made to correct the moisture condition leading to the damage. If the condition cannot be corrected, all infected parts should be replaced with treated wood or with all-heartwood lumber of a naturally decay-resistant wood species. If the sources of moisture that caused the decay are entirely eliminated, it is only necessary to replace the weakened wood with dry lumber.

Bacteria

Most wood that has been wet for any considerable length of time probably will contain bacteria. The sour smell of logs that have been held under water for several months, or of lumber cut from them, manifests bacterial action. Usually bacteria have little effect on wood properties, except over long periods of time, but some may make the wood excessively absorptive. This effect has been a problem in the sapwood of millwork cut from pine logs that have been stored in ponds. There also is evidence that bacteria developing in pine veneer bolts held under water or water spray may cause noticeable changes in the physical character of the veneer—including some strength loss. Additionally, mixtures of different bacteria, and probably fungi also, were found capable of accelerating decay of treated cooling tower slats and mine timbers.

SOURCE

U. S. Department of Agriculture, Forest Products Laboratory, *Wood Handbook: Wood as an Engineering Material,* Forest Service Agriculture Handbook No. 72, U. S. Government Printing Office, Washington, D.C., 1974.

BIOLOGICAL DETERIORATION OF WOOD

Kinds of Damage

There are four primary types of damage by wood-attacking fungi commonly recognized: Sap stain (chiefly dark, commonly known as blue stain), mold, decay, and soft rot. The distinctions between fungi types are useful but not always clearly defined. Bacteria will also degrade wood under special conditions.

Fungi destroy more wood than do any other organisms. Those causing decay, or rot, are by far the most destructive. In their simple growing stage they are threadlike, and the individual strands, called hyphae, are invisible to the naked eye except in mass. These hyphae penetrate and ramify within wood.

Botanically, the fungi are a low form of plant life. They have no chlorophyll; therefore, they cannot manufacture their own food but must depend on food, such as wood, already elaborated by green plants. Fungi convert wood they are invading into simple digestible products, and in the process the wood loses weight and strength.

Sap Stain

Sap stain is a discoloration that occurs mainly in logs and pulpwood during storage and in lumber during air drying. As its name implies, it is a discoloration of the sapwood. The "blue-stain" type, which dominates and is the only one of the sap stains of much importance commercially, is caused by the dark color of the invading fungus. Sap stain can go deep into the wood, causing a permanent blemish that cannot be surfaced off. The color of sap stain usually ranges from a brownish or a steel gray to almost black, depending on the fungus and the wood species.

Sap stain alone ordinarily does not seriously affect the strength of wood, but heavily stained wood is objectionable where strength is of prime importance. Its presence signifies that moisture and temperature have been suitable for the development of decay fungi; hence, early decay often is present, though masked by the stain. Stained wood is more permeable to rainwater; thus wood in exterior service is more subject to decay infection.

Mold

Molds cause discoloration that is largely superficial and can be removed by brushing or shallow planing. In coniferous woods the discoloration imparted by mold fungi typically is caused by the color of surface spore masses (green, black, orange); in hardwoods the wood itself often is superficially discolored by dark spots of various sizes.

Mold hyphae, however, penetrate wood deeply and increase permeability, sometimes very markedly. Also, heavy molding often is accompanied by hidden incipient decay.

Unpainted Pennsylvania Dutch barn that has withstood the danger of fungi, rot, and deterioration for many years. (Photo by F. Wilson)

Visual Inspection

Some indications in building components of moisture accumulation and decay are listed below:

1. Rust around nail
2. Surface mold on wood, and failure of paint on areas near absorptive end grain
3. Sunken surface or visibly deteriorated wood, often with paint failure
4. Fruit bodies of fungi (evidence of advanced decay)
5. Nail pulling (evidence of alternate shrinking and swelling)
6. The "pick test" for early decay: Wetted wood if sound lifts as a long sliver or breaks by splintering; if infected it tends to lift in short lengths and to break abruptly across the grain without splintering.

Recognition of Decay and Serious Wetting

Discoloration of the Wood

As decay progresses it usually imparts an abnormal color to wood. This change in color can be a useful diagnostic of decay if the inspector is reasonably familiar with the color or color shades of the sound wood. On surfaced wood the discoloration commonly shows as some shade of brown deeper than the sound wood. Some decays, however, produce a lighter than normal shade of brown, and this change may progress to a point where the surface might be called white or bleached. If this bleaching is accompanied by fine black lines (zone lines), decay is virtually certain. Often an abnormal variation in color creating a mottled appearance is more helpful in detecting early decay than actual hue or shade of discoloration. Highly indicative of decay, and expecially conspicuous, is variable bleaching on a dark background of blue stain or mold.

Accompanying the color change, there may be an absence of normal sheen on the surface of infected wood. Here also, familiarity with the normal appearance of the wood can be of great help in recognizing the loss of sheen. Occasionally, in relatively damp situations, the presence of decay infection will be denoted by surface growth of the attacking fungus; in these cases the wood beneath usually is weakened, at least superficially.

Stain showing through paint films, particularly on exterior woodwork, is evidence of serious wetting and probably decay beneath the film. Rust around nail heads suggests that wetting has been sufficient for decay to occur.

Loss of Wood Toughness and Hardness

Wood can also be examined for decay by simple tests for toughness of the fibers and for hardness. Toughness is the strength property most severely reduced by early decay. The pick test is a helpful and widely used simple means of detecting diminished toughness. It is made most reliably on wet wood. An ice pick, small chisel, sharpened screwdriver, or similar sharp-pointed or edged instrument of tough steel is jabbed a short distance into the wood and a sliver pried out of the surface. The resistance offered by the wet wood to prying and the character of the sliver when it finally breaks are indicative of toughness.

In the pick test, sound wood tends to break out as one or two relatively long slivers and the breaks are of a splintering type. Where loss of toughness has been appreciable, the wood tends to lift out with less than usual resistance and usually as two relatively short pieces. Moreover, these short pieces break brashly at points of fracture; that is, abruptly across the grain with virtually no small splinters protruding into the fracture zone.

On planed lumber, the reduced toughness of wood with early decay is sometimes indicated by abnormally rough or fibrous surfaces. Similarly, the end grain of a board or timber may be rougher than usual after sawing.

Toughness may also be reduced by certain other factors, such as compression wood, tension wood, or compression failures. There usually is little doubt of decay infection if the weakening is accompanied by a decay-induced type of discoloration.

In many cases the reduced hardness of infected wood can be detected by prodding the wood with a sharp tool. Softening, however, usually is not so obvious or so easily detectable with early decay as is a decrease in toughness.

Millbrook, N.Y. (Photo by F. Wilson)

Shrinkage and Collapse

Decay in the more advanced stages frequently causes wood to shrink and collapse. Under paint, this may first be manifested by a depression in the surface. Often the paint will acquire a brown-to-black discoloration from soluble materials migrating from the interior zone of decay to the outside; also, fruit bodies of the attacking fungus sometimes appear on the surface.

In exterior woodwork, wetting is often evidenced by brown-to-black discoloration or loosening of paint, particularly at joints. If there also has been substantial decay, the discoloration may be associated with interior collapse and surface depression.

Surface Growths

Decay in crawl spaces invaded by a water-conducting fungus may be evidenced by fanlike growths, vinelike strands, or by a sunken surface of wood resting on foundation walls or piers. Such decay is usually most advanced near the foundation because the fungus usually starts there. The fanlike growths are papery, of a dirty white with a yellow tinge. They may spread over the surface of moist wood, or more commonly, between subflooring and finish flooring or between joists and subfloor. These growths may further appear under carpets, in cupboards, or in other protected places. Water-conducting, vinelike strands grow over the foundation, framing, the underside of flooring, inside hollow concrete blocks, or in wall voids.

The fungus carries water through these strands from the damp ground or other source to the normally dry wood being attacked. Usually the main water conductors are ¼–½ in. wide, although they sometimes reach 2 in. They are similar in color to the fanlike growths, although they sometimes turn brown to black. During dry weather, shrinkage cracks in floors often outline the extent of an attack. Rotted joists and subflooring in relatively dry crawl spaces usually have a sound shell even when the interior wood is essentially destroyed.

REFERENCE

U. S. Department of Agriculture, Forest Service, Forest Products Laboratory, Madison, Wis., Research Paper FPL 190 (1973).

TEST METHODS FOR TIMBER

TEST METHOD	PROPERTY OR PARAMETER	CAPABILITY	ADVANTAGES	LIMITATIONS
Visual/ optical				
Visual Inspection	Extent of decay: Species	Visual inspection to search for characteristics of decay that are typical of different structural components such as siding, roofs, etc.	A good preliminary step in structural assessment to yield an overall evaluation	Inspection should be followed by other test methods to assess internal stability
Visual stress grading	Strength and grade	Examination of such qualities as sizes and frequency of knots, grain slope, and wane lead to a stress reduction factor (compared to stress value of clear wood)	Well suited for grading inspection. Provides a measure structural adequacy related to conventional practices under accepted ASTM/ALS grading practices	Limited to accessibility. May be impractical if grade mark or wood is covered with paint
Manual probing	Extent of decay	Pulling out of surface splinters of decayed wood and comparing with splintering and breaking characteristics of sound wood	A good detection method for surface decay. It is fast, easy, and decay characteristics are easy to identify (if they are in an advanced stage of development)	Inspection should be used in conjunction with another test method to assess internal quality. In existing buildings, not all surfaces may be accessible. Cannot measure decay, unless it proceeds inwards from the surface

TEST METHOD	PROPERTY OR PARAMETER	CAPABILITY	ADVANTAGES	LIMITATIONS
Penetration tests—Pilodyn penetrometer	Density (strength) and degree of degradation.	Can estimate approximate in-situ strength properties and degree of decay	Equipment is portable, simple can be used by field personnel with appropriate training	Does not provide a precise determination of strength. Readings must be calibrated with known samples. Cannot measure decay unless it proceeds inwards from the surface. Measures only advanced stages of decay.
Electrical				
Dielectric-type moisture meters	Moisture content	Capacitance meter: Measures a change in oscillation frequency due to moisture content/dielectric constant of the wood or change in the capacitance of the electrode as an impedance element when in contact with the specimen. Power-loss meter: Measures a loss of amplitude of an electrical wave emission resulting from amount of moisture in wood	Both types are easy to use. There is no physical disturbance of the surface.	Useful range of the dielectric-type moisture meters is from 0% to approximately 30% moisture content. Sensitive principally to the surface of the sample. Accuracy is relatively low, particularly when moisture gradient is present. Readings are affected by specimen density, chemical treatments, or decay
Resistance-type moisture meter	Moisture content	Moisture content of any size piece of lumber is determined by measuring its electrical resistance between two probes inserted in the lumber	Equipment is simple and rugged. Readout is in direct units, calibrations available for other grades and species	Yields approximate results in only the 7–30% moisture content range. The data are influenced substantially by some preservatives, fire retardants, and decay
Electrical resistance probe	Moisture content	Moisture content is measured by the electrical resistance between two electrode faces on a small wooden probe inserted into a test sample	Long-term moisture content changes can be measured by remote means. Can be built into the structure	Has only been used in research, therefore in-situ use is questionable. The probes often show long-term drift and hysteresis. Useful range is 7–35% moisture content

TEST METHODS FOR TIMBER (continued)

TEST METHOD	PROPERTY OR PARAMETER	CAPABILITY	ADVANTAGES	LIMITATIONS
Pulse velocity				
Ultrasonic and impact-induced pulse velocity equipment (longitudinal wave propagation)	Strength; modulus of elasticity	Major variations in the velocity of a longitudinal wave propagation indicate possible discontinuities. The transmission time for longitudinal wave propagation is measured and then translated into modulus of elasticity; this is used to estimate degree of decay and to assist in determining strength from tables of sound wood	Equipment is portable and readily adaptable for field use. Relatively fast measurements	Velocity can be affected by wood characteristics that are not flaws (such as moisture content), reducing accuracy
Stress-wave propagation equipment	Strength; modulus of elasticity	The propagation or velocity of transverse stress waves is influenced by inconsistencies in the wood that may affect strength. The density and wave velocity in a sample are measured to yield a modulus of elasticity. Strength evaluations are based on that value	Portable, light weight, low cost	Requires trained personnel to operate
Weight Test				
Oven- drying	Moisture content	Samples of wood are taken from a structural member and differentially weighed to determine moisture content (before and after drying)	Accurate results can be expected at any level of moisture content.	Requires lab test equipment. Takes considerable time
Radiographic evaluation	Grain direction; irregularities; decay; splits; knots; moisture content; insect damage; location and size of members in a floor or wall system	Capable of measuring thickness variations over 4% in ½-in.-thick materials (10% in 1-in.-thick materials). Can detect internal density variations.	Provides a permanent record. Equipment for wood evaluation is lightweight, portable. Test is easy to perform.	Radiation is harmful to organic tissue and must be shielded. High initial cost. Time delays for film development. Must have access to opposite sides of test specimen.

REFERENCE

Selected Methods of Condition Assessment of Structural, HVAC Plumbing and Electrical Systems in Existing Buildings: NBSIR 80-2171. Frank H. Lerchen, James H. Lichen, James H. Pielert, Thomas K. Faison.

TEST METHODS FOR TIMBER

Even though timber, like steel, is a mill product its in-place characteristics are generally more uncertain. To begin with, it is necessary to determine the species of wood before starting to estimate unit weight, tensile, compressive, and shear strengths, or its moduli. For example, unit weights among softwoods can range from about 20 pcf to over 37 pcf depending on species and, to a lesser extent, on moisture content. Even more radical is the range of some mechanical properties; for example, allowable fiber stresses in bending can range from 225 psi to almost 3500 psi depending upon softwood species, moisture content, and grade. Coefficients of variation for within-species clear wood may easily exceed 15% for modulus of rupture. Even the modulus of elasticity can range from 600,000 psi to 2,000,000 psi. Within-species variation of modulus of elasticity exceeds 20% for clear cut specimens.

The in-place moisture content can vary within a given member and even more so between members on different floors. Consider, for example, wood joists over a damp basement and those under a dry attic.

Grain patterns and knots can be extremely irregular and radically affect the strength of individual members although the impact of local irregularities on the strength of an assembly can be mitigated by their randomness.

Visual Inspection

Extent of Decay

Visual inspection and probing with a sharp instrument are usually the first and most comprehensive test methods employed when inspecting wood structural members. Decay is mainly the result of fungus associated with moisture accumulation. Characteristics of decay to be noted in different types of structural components are given in the following listing:

Component	*Characteristics of Decay*
Siding	Abnormal coloring (deeper than normal), brown color, and cubical checking indicates an advanced stage of decay. Bleaching with or without the presence of fine black lines, and softening especially where siding ends butt against trim or each other
Foundations	Fanlike fruiting bodies of fungi located between subfloor and finish floor, and between joists and subfloor

Roofs	Cubical checking, warping, softening, shredding, breakage. The location of all these deficiencies may be on the underside of roof sheathing
Porches	Same as siding. Concavely worn areas that may trap water should be examined especially.
Windows and doors	Brown or black discoloration near joints. Stains on sash from condensation. Softening and mold growth from accumulation of condensate

Advantages. Visual inspection is the preliminary step in evaluation of a member's strength properties. An estimate of the in-situ surface condition obtained in this manner gives a first estimate of the presence of decay. Such evidence suggests the possibility of serious loss of strength. Since this is a subjective test the accuracy of the results is dependent on the skill of the inspector. The results are not indicative of the structural capacity if there is any suspicion of internal decay.

Visual Stress Grading

Strength and Grade

Visual stress grading is a nondestructive method of identifying and categorizing lumber according to its defects, which can assist in assigning it to broad categories of anticipated load bearing capacity by:

1. The formulation of grading rules; and
2. The application of successive reduction factors to the strength values obtained from small clear tests to take into account the different strength-reducing influences involved. In this manner, an estimate can be made of the percentage of the test specimen's strength compared to the strength of perfect material of the same size and species, for example, 50 or 75%. The characteristics that determine a visual stress grade include size and frequency of knots, checks, splits, and pitch pockets; slope of grain; density; and wane.

A test specimen often will have several characteristics that can affect a particular strength property. Consequently, only the characteristic that gives the lowest strength ratio is used to derive the estimated strength.

When using the visual stress grading method, the modulus of elasticity assigned to a grade is an estimate of the mean modulus of the lumber grade. The average modulus of elasticity for a clear wood of the species is used as a base and can be obtained from such references as "Standard Methods for Establishing Clear Wood Strength Values," from the American Society for Testing and Materials. This clear wood modulus of elasticity is multiplied by empirically derived "quality factors" to obtain the reduction in modulus of elasticity that occurs by grade of lumber.

Advantages and Limitations. Visual stress grading provides an assessment of the conformance of lumber grades in structure to current practices. It permits an assignment of design values based on current practice. When applied to in-place lumber, the procedures are limited to accessible samples and may possibly deface the test pieces, for example, by damaging paint or other finish. When properly applied, visual stress grading yields an approximate, and generally conservative, estimate of a member's strength because there is little allowance made for specific gravity of the wood and only the surface is available for inspection.

It must be assumed that a defect has the maximum effect that can be deduced from a surface examination. The method has the limitation that it cannot differentiate the modulus of elasticity for different grades, which means that, where deflection is a limiting factor, there is little advantage given to the higher grades. Actual modulus of elasticity for individual pieces of the grade probably will vary from the mean assumed for design. The properties derived from visual grading criteria should be modified for design use by considering the influence of size, moisture content, and load duration. The test requires a trained grader or careful attention to both grading and stress assignment procedures. Both hardwoods and softwoods can be assessed with this technique, but many hardwoods will require interpretation for development of stress values.

Manual Probing

Extent of Decay

Prodding or probing a suspected piece of decayed wood with a sharp tool and observing the resistance to marring gives an idea of the stage of decay. A loss of hardness can be determined by comparison with sound wood of the same species. Sound wood tends to lift out of the stock as one or two long slivers with splintery breaks when jabbed with a pointed tool. Decayed wood tends to lift out and break off squarely across the grain with little splintering and little resistance.

Advantages and Limitations. Manual probing will yield an accurate assessment of in-place surface decay if it is extensive, but it must be used in conjunction with other test methods, as described in the table, to determine the internal quality. The probing method is best suited to framework, siding, and fences. Penetrometers often are used to determine the quality and uniformity as well as strength properties of wood.

Penetration Test—Pilodyn Nondestructive Wood Tester (penetration)

Degree of Degradation and Associated Reduction in Strength, Density, and Shock Resistance of Structural Wood

This test is based on the principle that the degree of penetration of a steel pin into wood is dependent largely on the nature and extent of decay present in the wood. It is assumed that the pin will penetrate deeper into decayed wood than into healthy wood. Research indicates an inverse proportionality between penetration (density) and amount of energy required for the penetration and a parabolic relationship between penetration (density) and pin diameter. Research further indicates that there is a correlation between density and strength of the material.

Some of the applications of this test method are:

1. Estimate the strength of biodeteriorated wood
2. Estimate the condition of wood foundations
3. Assess the degree of chemical or thermal decomposition on wood
4. Assess the residual strength of poles decaying from the outside
5. Estimate the density (strength) of sawn timber (stress grading of high-quality timber).

The test is conducted by loading the penetrometer device, pressing it firmly into the wood surface to be tested, and pulling the trigger to release the striker mechanism. The degree of penetration for a known constant energy and pin diameter then is measured by a scale reading from 0 to 40 mm (0 to 1½ in.). The instrument measures the fracture surface area created by a constant amount of energy. As an example, the "impact work" of a typical test might be of the order of 80 kj/m² (5,484 ft-lb/sq. ft).

The test should not be made on cracks, fissures, or areas of excessively rotted wood, since the results would not be indicative of the overall sample.

Advantages and Limitations. The equipment is portable and easy to use with limited amount of training, and its weight is only about 1.2 kg (2.6 lbs). It can be used either on unimpregnated or salt-impregnated wood. Unreliable values may result from the test if the wood is frozen or in an extremely dry condition.

The correlation between depth of penetration and the associated effective reduction in the strength of the structural wood member is affected to some degree by: (1) type of wood, (2) moisture content, and (3) normal variations in density. For this reason, it is advisable to calibrate the scale readings by means of comparative tests on known samples of wood.

The test is reliable if used under uniform moisture conditions. It is known that moisture content influences the penetration, but further research is necessary in order to establish a fully reliable relationship. Research has shown that this device is suited best to use on wood where the strength loss due to degradation occurs on or near the specimen surface.

Dielectric-Type Moisture Meter

Moisture Content

Capacitance-type moisture meters use the principle that the frequency of an oscillator changes according to the effect the specimen has on the circuit capacitance, or, in other words, according to the dielectric constant of the specimen. A frequency discriminator generates a signal, which can be read on a meter, that is proportional to the changes in frequency. Using the relationship between the dielectric constant and moisture, the meter then can be calibrated to read moisture content in-situ.

Radio-frequency power-loss moisture meters use the relationship between moisture content and power loss factor as a measure of moisture content. The wood specimen is penetrated by the electric field radiating from an electrode that is coupled to a low-power oscillator. The amplitude of the oscillation is indicated by a micro-ammeter. As the wood absorbs power from the oscillating field, the amplitude of oscillation is reduced, which results in reduced meter current. Because of the correlation between moisture and power-loss factor, the meter can be related to percentage moisture content. The calibration is empirically related to the average density of a given species of wood. Readings should be taken near the middle of the widest surface at least 500 mm (20 in.) from the end of the specimen.

Advantages and Limitations. Dielectric-type moisture meters are relatively easy to use. The useful range of the measurement of moisture content is from 0 to 25%. Measurements are less accurate, however, when moisture gradients exist in the wood since they are very sensitive to surface moisture. The accuracy of readings can be affected by: (1) calibration of the meter, (2) the species of wood being tested, (3) temperature, (4) operator skills, (5) chemicals in the wood, (6) specimen thickness, (7) moisture distribution, and other factors. Species and temperature correction factors must be applied if the species or temperature of the test sample are different from those of the material used for calibrating the meter. Treatment of wood with salt preservatives or fire retardants, or long-term exposure of the wood to seawater, will give unreliable meter readings.

Resistance-Type Moisture Meter

Moisture Content

Generally, resistance-type moisture meters are portable, battery operated, wide-range ohmmeters. The meters are of a direct reading type, calibrated for one species of wood with corrected tables provided for other species. For the resistance-type meter, penetration of the wood specimen by metal electrodes is necessary to measure its electrical resistance. The method of contact must be reproducible to provide consistent and meaningful results. The points of contact are pin-type electrodes that penetrate the surface of the specimen to a specified depth. The most common types of electrodes are: two-pin, four-pin, and insulated-pin. The insulated pin is useful because it allows testing of lumber that has a high superficial moisture content (such as is caused by rain or dew). One of the major differences between the electrode types is the depth of penetration. The four-pin penetrates approximately 8 mm ($\frac{5}{16}$ in.), the two-pin penetrates approximately 25 mm (1 in.), and the insulated pin penetrates approximately 25–76 mm (1–3 in.). Two nails may also be used for electrodes and calibration data for two-pin electrodes applied. Measurements on wood 6 mm ($\frac{1}{4}$ in.) or less in thickness are made with an electrode consisting of approximately 6–12 short, fine needles. Readings should be taken near the middle of the widest surface at least 500 mm (20 in.) from the end of the specimen.

Advantages and Limitations. The instruments are relatively simple and rugged, and are especially useful for in-situ evaluation of structural wood members. The useful range of the resistance-type moisture indicator is from approximately 7 to 30% moisture content with an accuracy of ± 10%. Electrolytes from preservatives or fire retardants can produce erroneous readings, and this can be a serious limitation to its use. It should be realized that meters read only the wettest material in contact with the electrode. Thus, an uninsulated pin-type test on wood with a high surface moisture content would give misleading readings. Species and temperature corrections must be applied if the species or temperature of the test sample are different from those of the material used for calibrating the meter. Needle-type electrodes must be used (surface contact electrodes are not usable). Insulated electrodes are preferred because of the possibility that unknown moisture gradients may exist in the wood.

Electrical Resistance Probe

Moisture Content

An electrical resistance probe has been developed by the U. S. Forest Products Laboratory using wooden elements as the moisture sensor. The principle is the same as for the other types of resistance measuring devices, but the probe construction is unique. The probe consists of a rectangular element of wood having two of its opposite faces coated with an electrically conductive silver paint. Two silver-painted electrodes are connected to a resistance-type moisture meter by two lead wires and measurements are taken based on the principle that the electrical resistance between the two electrode faces changes in response to change in moisture and temperature. The probe can be calibrated and located in remote areas and wired back to the meter for readout.

Advantages and Limitations. The wooden probe offers the advantages of reflecting the temperature and relative humidity conditions which influence the moisture content of the wood products to which it is placed. The probe will seek equilibrium of the material nearby and long-term moisture content changes can be indicated by remote means. It can operate over a moisture content range from 7 to 35%. The major application for this type of probe has been in research involving in-situ evaluation of structural wood members. These probes often show substantial long-term drift, hysteresis, and slow response to changing conditions.

Ultrasonic and Impact-Induced Longitudinal-Wave Pulse Velocity Equipment

Mechanical Strength, Defects and Decay, Density, Modulus of Elasticity

This test method operates on the principle that discontinuities (as well as other factors which affect modulus of elasticity) in wood will attenuate the velocity of longitudinal wave propagation through it. Major variations in the velocity at various time intervals would indicate irregularities in the wood such as knots, cracks, voids, etc. (This has not been quantified in practice.) The apparatus measures transmission time of longitudinal wave pulses from one piezoelectric transducer to another; both are attached to opposite surfaces of a piece of lumber. The frequency of the pulse is a function of the commercial equipment and usually varies between 150 and 1,000 kHz. (Frequencies above 500 kHz do not penetrate dry wood well.) The transmission time is translated into an estimate of strength and grade.

Advantages and Limitations. The properties derived from this test method should be modified for design by considering the influence of size, moisture content, and load duration. Certain defects have not been studied enough to know if they can be detected reliably using this test method. The equipment requires specially trained operators. Although it is most applicable as a production-type lumber scanner, it can be adapted for field use. It can sense defects in boards and laminated beans up to 51 mm (2 in.) thick and has a potential for locating surface and internal defects in lumber of all species, green or dry, rough-sawn or surfaced.

Stress-Wave Propagation Equipment

Strength, Modulus of Elasticity, Presence of Mechanical Flaws

This test method is based on the principle that the velocity of stress waves through a material depends on the density of the material and increases as density increases. Voids and decay in wood cause a reduction in the velocity as compared to the velocity in a healthy, denser, wood.

Stress-wave propagation testing is somewhat similar to the ultrasonic vibrational method in that it is a dynamic test, but the frequency of excitation is approximately 1,000 times higher than the sonic vibration methods. The modulus of elasticity E is calculated by measuring the period of time that it takes a stress wave to cover a given distance in the wood (wave velocity C), measuring the density p, and applying the following equation of approximate relationship:

$$E = pC^2.$$

The simplified test procedure is as follows: (1) A high frequency stress wave is induced on a wood sample by an excitor mechanism. (2) The propagation time of the stress wave is measured between two electron sensors that are a standard distance apart. (3) The piece of lumber is weighed to determine its density. (4) Test results are entered in the equation and a value of modulus of elasticity is determined. The modulus of elasticity then can be interpreted further into strength properties and grade of the wood.

The modulus of elasticity assigned to a grade is intended to be an average value for that grade. Stress grading machines can be adjusted so that the modulus of elasticity for a grade varies less than it would in a visual stress grade. It has been demonstrated that strength ratio and modulus of elasticity used together provide a more efficient prediction than either by itself. Two types of stress waves are useful for grading lumber, namely: (1) the longitudinal stress wave induced in the end of a piece of lumber, and (2) the flexural wave. Research on dry wood indicates that the longitudinal stress wave typically yields a higher estimate of E than the flexural stress wave. Although the various techniques yield somewhat different values of elasticity, the moduli determined with dry lumber have been shown to correlate well.

Advantages and Limitations. Propagation of stress waves is influenced by the mechanical and physical properties of the timber sample and somewhat by the shape of the piece. The cited equation for stress-wave analysis generally assumes the medium to be isotropic and homogeneous; this is misleading, mainly because of natural inconsistencies in the wood. It has been found that the modulus of elasticity decreases with increasing moisture content up to fiber saturation, which is about a 25% moisture content. Thus, a false value of elastic modulus could be evident for data taken on wood above fiber saturation. It is necessary to consider this in interpreting results, since moist wood could produce a short stress-wave time, even though voids or decay might be present. The electronic measuring equipment is very sensitive and has been refined to the point of measuring only the wave movement pertinent to the needed results. The properties derived from mechanical stress grading criteria should be modified for design by considering the influence of size, moisture content, and load duration. Modulus of elasticity calculated from the basic longitudinal stress-wave formula is much more dependent on moisture content than E calculated from the basic flexural stress-wave formula. The longitudinal stress-wave method of stress grading may overestimate the static E of incompletely dried lumber. More importantly, when used to grade on the basis of stress-wave speed over short spans, it may cause dry lumber containing wet pockets to be downgraded. Flexural-wave stress grading may also be affected by wet pockets, depending on their position as well as their moisture content. The device can be used in relation to either calibrated standards or a heavy degree of technical interpretation. Lightweight, low-cost units are available. The test usually requires two persons to operate the equipment. One of them should be knowledgeable about the pathology of decay in wood and should understand the stress wave technology well enough to know when the readings are appropriate to the known conditions and be able to determine the properties of normal wood. Although this test is most applicable to production-type lumber scanning, it can be adapted for field use.

Oven-Dry Test

Moisture Content

For this test method, the moisture content of a wood sample is calculated from weight values obtained before and after drying it in an oven which is maintained at $103 \pm 2°C$ ($217.4 \pm 3.6°F$).

To conduct the test, a full-cross-section specimen is cut or core drilled from the host sample so that it is no less than 25 mm (1 in.) along the grain, or longer as needed to provide a minimum volume of 33 cm³ (2 in.³). The specimen is weighed immediately after cutting (or protected in a vapor-tight container until it can be weighed). Weighing must be to an accuracy of ± 0.2%. The maximum delay for weighing after cutting is 2 hours. Then the specimen is dried in the oven until it reaches a constant weight and is weighed again after removal from the oven. The moisture content then is calculated in percent as $[(A - B)/B] \times 100$, where A is the original weight and B is the oven-dried weight.

Advantages and Limitations The greatest advantage to this test method for measuring moisture content in wood is its accuracy and time-tested reliability. Accurate results can be expected at all levels of moisture content. The major disadvantage is that it must be performed in a testing laboratory (not on-site).

Radiography

Grain Direction, Internal Discontinuities, Moisture Content

Radiography is the imaging of a radiation beam that has passed through an object under study. The test is based on the principle that radiation attenuation is caused by density variations in the test specimen. In the case of wood, it can be used effectively to determine changes in grain direction (as around knots and other irregularities), to detect internal discontinuities and gross density variations (such as mineral streaks, decay, insect damage, splits, etc.), and to detect changes in moisture content. It also can be used for mapping studs, joists, bracing, posts, and other structural elements of a wood frame building for the purpose of determining the strength of wall or floor systems.

To conduct the test, x-ray radiation is emitted from a source (e.g., x-ray tube, radioactive isotope, etc.), penetrates the specimen, and exits the opposite side of the specimen where it is recorded on a sensitive film as a radiograph (picture) or viewed on an electro-optical imager (screen). Voids or relatively low-density areas of the specimen allow a relatively higher rate of passage of the radiation which will appear as dark indications on the film (as compared to the relatively lighter background). Areas of higher density will provide the opposite effect on the film.

Generally speaking, the energy level used for x-radiation should be the lowest consistent with a reasonable exposure time. This allows detection of small changes in the test material thickness or density.

Advantages and Limitations. Radiographic evaluation techniques offer the advantage of providing a permanent record of the condition of wood specimens on film or a "live" image on an electro-optical system for greater speed of evaluation. Equipment for x-ray inspection of wood is available which is lightweight and readily portable. The test is relatively easy to perform. Radiographic sources are harmful to organic tissue, so the test environment must be shielded (or isolated) as necessary to prevent radiation hazards. Drawbacks to the use of radiographic evaluation equipment are the high initial cost of the equipment and the film, the time delay necessary to develop the film, and the requirement that access to both sides of the test specimen is necessary to set up and use the equipment.

For inspection material thicknesses up to 12.7 mm (½ in.), thickness variations of about 4% are detectable. But, for thicknesses of 25.4 mm (1 in.) or more thickness variations of 10% or more are detectable.

DECAY RESISTANCE OF WOOD

Wood kept constantly dry does not decay. Further, if it is kept constantly submerged in water, even for long periods of time, is not decayed significantly by the common decay fungi regardless of the wood species or the presence of sapwood. Bacteria and certain soft-rot fungi can attach submerged wood but the resulting deterioration is very slow.

A large proportion of wood in use is kept so dry at all times that it lasts indefinitely. Moisture and temperature, which vary greatly with local conditions, are the principal factors affecting rate of decay. When exposed to conditions that favor decay, wood deteriorates more rapidly in warm, humid areas than in cool and dry areas. High altitudes, as a rule, are less favorable to decay than low altitudes because the average temperatures are lower and the growing seasons for fungi, which cause decay, are shorter.

The heartwoods of some common native species of wood have varying degrees of natural decay resistance. Untreated sapwood of substantially all species has low resistance to decay and usually has a short service life under decay-producing conditions. The decay resistance of heartwood is greatly affected by differences in the preservative qualities of the wood extractives, the attacking fungus, and the conditions of exposure. Considerable difference in service life may be obtained from pieces of wood cut from the same species, or even from the same tree, and used under apparently similar conditions. There are further complications because, in a few species, such as the spruces and true firs (not Douglas fir), heartwood and sapwood are so similar in color that they cannot be easily distinguished. Marketable sizes of some species such as southern pine and baldcypress are increasingly taken from second growth and contain a high percentage of sapwood.

Precise ratings of decay resistance of heartwood of different species are not possible because of differences within species and the variety of service conditions to which wood is exposed. However, broad groupings of many of the native species, based on service records, laboratory tests, and general experience, are helpful in choosing heartwood for use under conditions favorable to decay (see following Tables). The extent of variations in decay resistance of individual trees or wood samples of a species is much greater for most of the more resistant species than for the slightly or nonresistant species.

Where decay hazards exist, heartwood of species in the resistant and very resistant categories generally give satisfactory service, but heartwood of species in the other two categories will usually require some form of preservative treatment. For mild decay conditions, a simple preservative treatment—such as a short soak in preservative after all cutting and boring operations are complete—will be adequate for wood low in decay resistance. For more severe decay hazards, pressure treatments are often required; even very decay-resistant species may require preservative treatment for important structural or other uses where failure would endanger life or require expensive repairs.

Grouping of Domestic Woods According to Heartwood Decay Resistance

VERY RESISTANT	RESISTANT	MODERATELY RESISTANT	SLIGHTLY OR NONRESISTANT
Locust, black	Baldcypress (old	Baldcypress (young	Alder
Mulberry, red	growth)*	growth)*	Ashes
Osage orange	Catalpa	Douglas-fir	Aspens
Yew, Pacific	Cedars	Honeylocust	Basswood
	Cherry, black	Larch, western	Beech
	Chestnut	Oak, swamp chestnut	Birches
	Cypress, Arizona	Pine, eastern white*	Buckeye
	Junipers	Southern pine:	Butternut
	Mesquite	Longleaf*	Cottonwood
	Oak:	Slash*	Elms
	Bur	Tamarack	Hackberry
	Chestnut		Hemlocks
	Gambel		Hickories
	Oregon white		Magnolia
	Post		Maples
	White		Oak (red and black species)
	Redwood		Pines (other than longleaf, slash, and eastern white)
	Sassafras		
	Walnut, black		Poplars
			Spruces
			Sweetgum
			True firs (western and eastern)
			Willows
			Yellow-poplar

* The southern and eastern pines and baldcypress are now largely second growth with a large proportion of sapwood. Consequently, substantial quantities of heartwood lumber of these species are not available.

Grouping of Imported Woods According to Approximate Relative Heartwood Decay Resistance

RESISTANT TO VERY RESISTANT	MODERATELY RESISTANT	SLIGHTLY OR NONRESISTANT
Angelique	Andiroba*	Balsa
Apamate	Apitong*	Banak
Brazilian rosewood	Avodire	Cativo
Caribbean pine	Capirona	Ceiba
Courbaril	European walnut	Jelutong
Encino	Gola	Limba
Goncalo alves	Khaya	Lupuna
Greenheart	Laurel	Mahogany, Philippine:
Guijo	Mahogany, Philippine:	Mayapis
Jarrah	Almon	White lauan
Kapur	Bagtikan	Obeche
Karri	Red Lauan	Parana pine
Kokrodua (Afrormosia)	Tanguile	Ramin
Lapacho	Ocote pine	Sande
Lignum vitae	Palosapis	Virola
Mahogany, American	Sapele	
Meranti*		
Peroba de campos		
Primavera		
Santa Maria		
Spanish-cedar		
Teak		

* More than one species included, some of which may vary in resistance from that indicated.
SOURCE: U.S. Department of Agriculture, Forest Service, Forest Products Laboratory, Madison, Wis.

VI. Metals

Ultrasonic Pulse Velocity

Radiographic Evaluation (X-Ray)

Eddy Current

Magnetic Particle

Coupon Test

IRON AND STEEL

The properties of various irons and steels are closely related to their chemical compositions and grain structure, that is, the size and arrangement of the crystalline microscopic particles making up the metal. In the initial crude refining processes, the most important factor influencing grain structure is the chemical reaction of the carbon with iron as it cools.

Further refining processes, alloying elements, heat treatments and hot and cold working operations also affect grain structure. In ordinary steels, carbon primarily as well as other residual elements are important. In higher grades of steel alloying elements contribute special properties to the metal.

The essential physical property that distinguishes steel from iron is its malleability as it comes from the furnace and is initially cast. Other unique properties of steel are its ability to resist high stresses by deforming without breaking, its toughness, and its hardenability by hot and cold working.

Pennsylvania Station, Newark, N.J. (Photo by F. Wilson)

Beam connection in Newark Station. (Photo by F. Wilson)

In casting pig iron, only slight variations in properties occur as a result of the reaction between carbon and iron, and in the resulting grain structure of the metal. Most ordinary white and gray cast irons, therefore, like pig iron, are hard and brittle, have relatively high compressive and low tensile strengths.

However, with the proper combination of selected pig irons, scrap metals, and alloying elements, cast iron can be made with increased hardness, toughness, and corrosion and wear resistance. Casting also permits the formation of complex and massive shapes not readily produced by machining or rolling.

Plumbers' drainage, vent, and waste pipe, ornamental railings, and lamp posts are familiar cast iron products. In sanitary ware involving enameling and baking at elevated temperatures, cast iron's resistance to warping and cracking at high heat is an important advantage.

Malleable cast iron has improved breakage resistance as a result of increasing toughness and ductility, and is used extensively for small articles of builder's hardware. Special alloy cast irons with alloying elements of 1–5% and high-strength cast irons containing molybdenum, chromium, and nickel are also produced for special uses. Nodular cast iron, produced in the U. S. by the addition of magnesium, has greater strength, ductility, toughness, machinability and corrosion resistance.

Carbon Steels

Carbon steels are those in which the residual elements such as carbon, manganese, phosphorous, sulfur, and silicon, are controlled, but generally no alloying elements are added to achieve special properties. They contain up to 1.20%

(Photo by F. Wilson)

carbon, and other elements are controlled within specified limits or ranges. Over 90% of the steels manufactured into finished products are carbon steels.

Carbon steels are used for most products in the building industry, from structural shapes, concrete reinforcing bars, sheets, plates, and pipes to the smallest items of builder hardware. Carbon steels generally possess adequate strength, hardness, stiffness, malleability, and weldability, but poor resistance to corrosion.

Occasionally some copper and other elements may be added to improve atmospheric corrosion resistance, and by varying the chemical compositions, heat treatments and subsequent hot and cold working operations, other desirable properties can be enhanced.

Alloy Steels

Steel is classified as alloy steel when the content of alloying elements exceeds certain limits or when a definite range of alloying elements is specified. The term does not cover certain classifications such as stainless steel, tool steel, specialty steel, and others, although these steels do indeed contain specified amounts of alloying elements.

Alloy steels are generally grouped by the element or combination of elements used to obtain desired properties. Manganese-molybdenum steels, for example, are known for hardenability and resistance to fatigue; silicon-manganese steels for high relative strength and shock resistance; nickel steels for toughness and impact and corrosion resistance.

(Photo by Hank Tenny, courtesy of American Iron and Steel Institutes)

(Photo courtesy of American Iron and Steel Institute)

The alloy steels have properties essentially similar to or exceeding those of carbon steels. In general, particular properties are improved by certain elements. For example, high silicon content gives steel excellent magnetic permeability and low core loss, making it useful in motors, generators and transformers; cobalt improves magnetic properties for such uses as permanent magnets in electrical apparatus and sealing gaskets in modern refrigerators; nickel imparts great toughness, useful in rock drilling and air hammer equipment.

Because of their higher cost, alloy steels are not generally used in construction materials. However, a particular group known as heat-treated constructional alloy steels have been used successfully in large structures where unusually high loads or temperatures are encountered; these steels are rolled in a limited variety of structural shapes. They are capable of developing very high strengths (ultimate strength up to 135,000 psi, yield point up to 100,000 psi) and show promise of increased structural use.

Alloy steels are used extensively in the manufacture of automobile and locomotive parts, excavating and heavy construction equipment, and for various industrial uses. Parts of common household equipment such as vacuum cleaners, washing machines, lawn mowers, and tools are sometimes made of alloy steels.

High-Strength Low-Alloy Steel

High-strength low-alloy steels comprise a group with chemical compositions specially developed to impart better mechanical properties and greater resistance to atmospheric corrosion than are obtainable from conventional carbon structural steels.

These steels are commonly used where high strength in relation to weight is important. They are often used in bridge and high-rise building construction. They are used wherever low maintenance against corrosion and impact damage is important.

Certain proprietary A242 steels can be made with increased corrosion resistance, when intended for exposed architectural use. These steels form natural, rust-colored, self-healing oxide coatings which inhibit further corrosion when left unpainted.

Stainless and Heat-Resisting Steel

Stainless steels possess excellent corrosion resistance at varying temperature ranges. Heat resisting steels retain their essential physical and mechanical properties at elevated temperatures. These two types of steel account for about 1% of steel production.

Chromium is the alloying element mainly responsible for corrosion-resisting and heat-resisting properties, although other elements such as nickel, manganese and molybdenum also contribute to these properties. The stainless qualities of steel are derived from a self-healing chromium oxide which forms a transparent skin and prevents further oxidation. Steels with less than 4% chromium are

(Photo by Jim Rankin, courtesy of American Iron and Steel Institute)

regarded as alloy steels. Heat-resisting steels normally contain 4 –12% chromium, and stainless steels over 12%.

Methods of Working

Casting

Castings are generally made by pouring molten steel into sand molds. Prior to casting pig iron may be further refined and melted with scrap metal and ferroalloys at the foundry to obtain desired casting and mechanical properties. Casting is used when desired shapes are of a size or complexity not readily obtained by rolling or machining. Complex and massive shapes possessing great strength and impact resistance at high and low temperatures usually are cast.

Wrought Steel

A small percentage of steel products, complex in shape and of irregular cross section, are manufactured by forging. Forging is a method of forming hot metal into desired shapes by pressing between heat-resistant dies. When metal is forged, toughness, strength, and ductility increase significantly.

Extruding

Semi-finished shapes can be converted into lengths of uniform cross section by extruding. In this process an advancing ram forces preheated, plastic metal through a tough, heat-resistant die of the desired profile.

Extruding produces more complex sections with better surface characteristics than are produced by rolling. It can be used to shape certain alloys without undesirable residual effects such as excessive hardness, and is more economical than other forming methods for small quantities.

Hot Rolling and Cold Finishing

By far the largest proportion of wrought steel products is manufactured by hot rolling and cold finishing. The hot steel passes through a system of rolls which gradually imparts rectangular bloom, billet, or slab shapes. These are cooled, cleaned of surface irregularities, inspected, and reheated for further rolling. Semi-finished shapes may be hot rolled directly into finished hot rolled products such as structural shapes, plates, sheets strip and bars; or hot rolling may be used as an intermediate step prior to cold finishing, as in the manufacture of bar, sheet, strip, and wire.

Hot-rolled products intended for cold finishing are descaled, cleaned of surface oxide scale.

Cold finishing consists of cold rolling, cold reduction, and cold drawing of

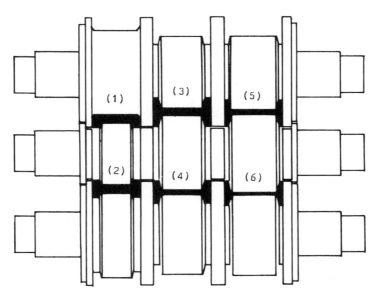

Steel rollers in rolling mill. (From Structural Systems, Cowan and Wilson, Van Nostrand Reinhold, New York, 1981)

previously hot-rolled, descaled shapes. Cold rolling involves passing metal at room temperature through sets of rolls to impart desired shape finish and mechanical properties. It improves the strength, surface finish, and flatness. Cold reduction may harden the metal excessively, so that it must be annealed to soften it and further tempered to obtain the proper strength and stiffness.

Cold drawing is used to make smaller or more complex bar shapes from hot rolled annealed bars, by pulling the bars through a hard, abrasion-resistant die. Wire is cold drawn from wire rods. This operation results in improved machinability and strength, smoother finish, and greater dimensional accuracy.

Structural Shapes

Structural members such as I-beams, H-sections, angles, channels, tees, zees, and piling are hot rolled from blooms or billets. As many as 20–30 passes through grooved rolls may be required to produce the desired shape. The first sets of rolls drastically reduce the cross section and impart the approximate shape; the finishing rolls gradually form the final shape and, for deep sections such as channels and angles, bend the legs into position. Wide flange shapes such as H-sections are rolled on universal mills, which have vertical rolls capable of shaping the vertical sections.

Protective Coatings and Mechanical Finishes

Ordinary cast iron, most carbon steels, and some alloy steels oxidize under ordinary atmospheric conditions. Elevated temperatures and moisture accelerate this oxidation, known as atmospheric corrosion. The resulting surface scale (rust) deteriorates the metal surface, and if not prevented progresses until the metal reverts to the oxide state found in nature. The chief function of protective coatings used on iron and carbon steel is to inhibit corrosion.

Not all ferrous metals require equal protection against progressive corrosion. Copper-bearing steels, some cast irons, and some high-strength low-alloy steels acquire excellent corrosion resistance from a tight surface oxide which inhibits further corrosion. Sometimes the appearance of the natural patina is objectionable and finish coatings are applied on these steels for decorative purposes. Decorative and protective coatings are used less often on stainless steels because of their superior corrosion resistance and attractive natural finishes.

Finishes for iron and steel may be classified as follows: (1) mechanical, including the as-rolled and as-drawn natural mill finish, and those imparted by further grinding, polishing or patterning; (2) chemical, generally consisting of cleaning or preparing operations for further finishing; and (3) organic and inorganic coatings, such as metallic, vitreous, laminated, and painted finishes. Organic coatings are commonly sprayed on or brush applied in the field.

REFERENCE

Olin, H., Schmidt, J., and Lewis, W., *Construction: Principles, Materials and Methods,* The Institute of Financial Education, Chicago, Illinois, and Interstate Printers and Publishers, Danville, Illinois, 4th Edition, 1980.

Sheet metal corner capital, Plainfield, N.J. (Photo by F. Wilson)

Welding faulty panel connection, Pan Am Building, New York City. (Photo by F. Wilson)

ALUMINUM

Corrosion Resistance

Aluminum's high resistance to corrosion is due to the very thin inert surface film of aluminum oxide which forms rapidly and naturally in air. After reaching a thickness of about a ten-millionth of an inch, this coating effectively halts further atmospheric oxidation of the metal and thus protects the surface. If this oxide film is broken, as by a scratch, a new protective film forms immediately. The oxide film increases in thickness with temperature and remains protective to the underlying aluminum even at the melting point.

It may be necessary to protect aluminum in certain severely corrosive environments by protective coatings such as organic paints, by cladding, or by increasing the thickness and effectiveness of the oxide film by anodizing.

Weathering

The extent of corrosion due to weathering depends on the type and extent of contamination of the surrounding atmosphere. More corrosive elements are present in industrial or marine areas than in inland rural areas.

In tests and observations, in part under the direction of the National Bureau of Standards, made over a 10-year period in different atmospheric conditions on various alloys, the depth of penetration of atmospheric corrosion ranged from less than 1 mil for rural areas, to about 4 mils for industrial areas, and up to 6 mils for seacoast locations.

Galvanic Corrosion

When two dissimilar metals are connected by a solution which conducts electricity (an electrolyte) galvanic corrosion may occur. As in a battery, an electric current is created, and one metal corrodes away while the other is plated. Galvanic attack can occur when moisture condenses from the air and contaminating elements act as an electrolyte. The threat of such corrosion is greatest in industrial areas because the atmospheric moisture is more contaminated, and in seacoast areas because of salt and moisture in the air.

The severity of galvanic attack varies depending on how far apart the two metals are from each other on the galvanic table. For example, coupling mild steel to aluminum would produce less corrosion than would coupling copper to aluminum under identical exposure conditions. The metal that is higher in the potential table is sacrificial and corrodes. Thus zinc provided cathodic protection to aluminum and aluminum alloys. That is, when zinc and aluminum are in the presence of an electrolyte, the zinc usually corrodes. Thus, zinc-plated steel parts are commonly used with aluminum to prevent the harmful effect of galvanic action on the aluminum. Cadmium-plated steel in contact with aluminum alloys is also satisfactory. There is no appreciable difference in the electrolytic potential of these two metals and hence practically no galvanic action.

Aluminum is compatible with some stainless steel alloys, chromium, zinc, and small areas of white bronze. Where permanent contact with other metals cannot be avoided, the risk of galvanic corrosion can be greatly reduced by painting the other metal and aluminum at the contact area with zinc chromate followed by one coat of a lead-free paint such as aluminum paint, or with a heavy-bodied bituminous paint. When the dissimilar metal cannot be painted, the aluminum may be given the insulating treatment. As an alternative, a strip of plastic or a similar insulator may be used in place of paint.

In severely corrosive atmospheres and high-moisture areas the edges of a dissimilar-metal joint may be sealed with compatible building mastic or caulking.

A dangerous source of electrolytic attack may be water drainage from metals such as copper. Therefore aluminum gutters on buildings with copper flashing must be protected by painting either the copper flashing or both the copper and aluminum.

Direct Chemical Attack

Direct chemical attack is the tendency of the chemical to dissolve the metal. Solutions of strong alkalis, sulfuric acid, hydrochloric acid, carbonates, and fluorides tend to attack aluminum because they can dissolve or penetrate the protective oxide coating. Some chemical attack may occur to aluminum in contact with wet alkaline materials such as mortar, concrete, and plaster, resulting in a mild etching of the aluminum surfaces. Hence surface protection to aluminum products may be necessary during construction.

Concrete and Masonry Contact

Where aluminum is buried in concrete, it is usually not necessary to paint the parts. However, if chlorides are present from additives or other sources in steel reinforced concrete, the aluminum parts should be coated with bituminous paint or other suitable coating. When in contact with concrete, masonry or other absorbent materials under wet or intermittently wet conditions, aluminum should be protected with a coating of bituminous paint, zinc chromate primer, or a separating layer of plastic or other gasketing material. Creosote and tar coatings should not be used because of their acid contents. Where wet concrete or other alkaline material may splash against aluminum during construction, the aluminum surface may be protected by a coating of clear acrylic-type lacquer or a suitable strippable coating if maintenance of appearance is important.

Although aluminum is inert when buried in many types of soil, some soils and the presence of stray electrical currents in the ground may cause corrosion. Buried aluminum pipe may have to be coated.

REFERENCE

Olin, H., Schmidt, J., and Lewis, W., *Construction: Principles, Materials and Methods,* The Institute of Financial Education, Chicago, Illinois, and Interstate Printers and Publishers, Danville, Illinois, 4th Edition, 1980.

TEST METHODS FOR METAL

Iron and Steel

Because iron and steel are mill-controlled products, far less variation can be expected in their in-situ characteristics than in cast-in-place concrete. This does not mean, however, that important variations cannot occur. This is particularly true of cast iron columns, which were very popular with architects around the turn of this century. The common deficiencies of these were principally (1) variations in shell thickness because of eccentric placement of the core at casting, (2) presence of blowholes and cinders in the shell, and (3) intrinsic variations in crystal structure caused by nonuniform cooling.

Steel shapes delivered to the site and erected can also have undesirable geometric variations in cross-section, camber, sweep, out-of-squareness, warping and length. Even though modern mill standards are far higher now than ever, post-mill (especially on-site) handling can create in-place members with far from assumed properties.

Further, erectors often deal with on-site geometric problems such as inaccurately placed column bases, out-of-plumb members, distorted members or seatings, poorly aligned rivet or bolt holes, etc., by brute-force methods, leaving more than a little residual strain in the completed structure.

When the characteristics of the steel are unknown, the following guidelines for yield strength may be used for existing structures:

1. Construction prior to 1905, f_y = 25,000 psi
2. Construction 1905–1932, f_y = 30,000 psi
3. Construction 1933–1963, f_y = 33,000 psi
4. Construction after 1963 f_y = 36,000 psi.

For wrought iron members the value may be assumed to be 25,000 psi.

Other Metals

Unlike bridge structures which are subjected to cyclic loadings, buildings generally are subjected to static loads (except for such conditions as seismic disturbances and vibrational loadings due to mechanical equipment). Consequently, flaws (such as cracks) in structural steel components of the building generally are not subject to progression from fatigue, and it may not always be critical to evaluate them. If, however, it is suspected that a structural component (a weldment, a splice plate, etc.) is degrading the structural integrity of the building, nondestructive evaluation (NDE) methods can be extremely useful to evaluate the component. On the other hand, it is important to evaluate flaws other than cracks (such as corrosion, voids, pits, fabricating discontinuities, porosity, etc.).

Most NDE methods offer more than just a superficial examination of the surface conditions of metals. It now is possible to discover facts concerning overall properties, quality, and dimensions of both the surface and the internal regions of the many metal test specimens.

Although nondestructive evaluation is used primarily on structural steel, some of the evaluation methods available are applicable to other types of ornamental and structural metals (e.g., cast-iron, aluminum, etc.) as indicated in the table under the column heading "Typical Applications." In all cases, the following list of factors should be considered in selecting the appropriate test method to use:

1. What is the material to be tested?
   (a) Is it magnetic or nonmagnetic?
   (b) Is it electrically conductive?
   (c) Does the metal have a nonconductive or nonmagnetic coating?
2. By what method was the metal fabricated (cast, wrought, powder metallurgy, welded, soldered, etc.)?
3. What is the geometry of the metal (surface condition, thickness, shape, etc.)?
4. What types of defects are possible or expected in the metal?
5. What degree of sensitivity and resolution is required from the test equipment?
6. What costs are involved?
   (a) How much does the equipment cost?
   (b) What are the operating and material costs of conducting the test?
7. How accessible is the metal (e.g., behind a wall, pipe, or wiring, concealed by ductwork)?

Test Methods for Metals

(This table indicates the most commonly used NDE methods for testing metals, cites examples of typical applications with their advantages and limitations, and describes the appropriate materials on which the test is used. A more detailed description of several of the most commonly used NDE methods follows.)

Test Methods for Metals

METHOD	PRINCIPLE OF OPERATION	PROPERTIES SENSED OR MEASURED	DEFECTS DETECTED	TYPICAL APPLICATIONS	ADVANTAGES	LIMITATIONS
Visual/optical	Special devices (borescopes, fiber optics, panoramic cameras, etc.) can be used to examine surfaces inaccessible to the naked eye. Magnifiers can be used to detect flaws too small to be seen by the naked eye	Material characteristics open to a surface	Surface flaws (cracks, voids, holes, gouges, fabricating discontinuities, corrosion, pits and other irregularities)	Surfaces of all metals	Permits examination of hidden surfaces (if available)	Detects only defects visible to the eye. Limited to detection of surface flaws only.

Test Methods for Metals (continued)

METHOD	PRINCIPLE OF OPERATION	PROPERTIES SENSED OR MEASURED	DEFECTS DETECTED	TYPICAL APPLICATIONS	ADVANTAGES	LIMITATIONS
Liquid penetrant	Liquid penetrant containing dye is drawn into surface defects by capillary action	Material separations open to a surface	Surface cracks, laminations, poor bond, gouges, porosity, laps, seams, stress cracks, fabricating discontinuities	Used on nonmagnetic metals. Used with casting, forgings, weldments, and components subject to fatigue or stress-corrosion cracking	Allows inspection of complex shapes in one single operations; inexpensive, easy to apply, portable	Will detect only defects open to the surface. It is messy. Irrelevant indications occur. Results are dependent on operator's ability to interpret results. Temperature of specimen, penetrant drain time, emulsifier soak and drain time, drying temperature, and developing powder dwell time must be controlled carefully to get true indications.
Ultrasonic	Vibrations above 20,000 Hz are introduced into metal sample Waves are reflected or scattered by discontinuities	Anomalies in acoustic impedance	Cracks, voids, porosity, laps, segregated inclusions, poor brazing or bonding. Will detect both surface and subsurface defects	Thickness gaging; material inspection of castings, forgings, and extrusions; for all metals	Locates small discontinuities; portable; instant results; accurate measure of thickness	Sensitivity is reduced by rough-surfaced parts. Odd-shaped pieces are hard to analyze. Requires skilled operator. Depends on operator's ability to interpret results and on orientation of the defect. Must couple transducer to surface of specimen carefully

Test Methods for Metals (continued)

METHOD	PRINCIPLE OF OPERATION	PROPERTIES SENSED OR MEASURED	DEFECTS DETECTED	TYPICAL APPLICATIONS	ADVANTAGES	LIMITATIONS
Magnetic particle	Magnetic particles are attracted to magnetic lines of leakage force and where breaks in the lines of force occur	Anomalies in magnetic field flux at surface of test sample	Cracks, seams, laps, voids, porosity, and inclusions	Surface and slightly subsurface inspection of parts sensitive to magnetization	Simple, inexpensive, senses flaws down to ¼ in. below surface as well as surface flaws	Not applicable to nonmagnetic metals or materials. It is messy. Careful surface preparation is required. Irrelevant indications often occur. Depends on the operator's ability to interpret results. Demagnetization after inspection may be necessary
X-ray gamma ray	The attenuation of x-rays is affected by the density of the test specimen. Voids or low-density areas show as dark indications on the x-ray film	Inhomogeneities in thickness, density, or composition	Voids, porosity, inclusions, and cracks	Used on castings, forgings, weldments, and assemblies to check for fatigue, thickness gaging, internal flaws, etc.; for all metals	Detects both internal and external flaws; portable; provides a permanent record on x-ray film	High cost; heavy; health hazard; not sensitive to defects less than about 2% of the total thickness of the specimen. Complex shapes are difficult to analyze
Eddy current	The impedance of a probe coil is measured constantly. The coil is placed in contact with the metal specimen. The coil impedance changes in direct relationship with the specific material properties and constituent variations	Anomalies in electric conductivity and, in cases, magnetic permeability	Surface finish discontinuities, dimensions, cracks, seams, variations in alloy composition or heat treatment	Used to evaluate condition of wire, tubing, local regions of sheet metal and alloy sorting; for thickness gaging; for electrically conductive or magnetically permeable metals	Moderate cost; readily automated; portable; permanent record available if needed; can be adapted to many comparative analyses	Useful on conductive materials only; shallow penetration; reference standards often are necessary; no absolute measurement—only qualitative comparison
Coupon	Stress-strain relationship; tension or compression tests	Stress-strain	Yield strength, yield point, tensile strength, elongation modulus of elasticity, compressive strength	Die castings, forgings, structural shapes, malleable iron, powdered metals	Gives fast, accurate results of physical and mechanical values	Test is destructive, since a sample must be removed to be tested

REFERENCE

Selected Methods for Condition Assessment of Structural, HVAC Plumbing and Electrical Systems in Existing Buildings; NBSIR 80-2171. Frank H. Lerchen, James H. Pielert, Thomas K. Faison.

Ultrasonic Pulse Velocity

Internal Flaws (Depth of Cracks and Their Rate of Propagation, Nonbonds, Inclusions, Corrosion, Interfaces, etc.)

Detecting flaws in metal by the use of ultrasonics is based on the principle that the velocity and amplitude of vibrational waves of acoustic energy propagated through a test specimen are affected by the elastic properties of the material being tested. To conduct the test, an ultrasonic pulse is generated by an electroacoustic transmitting transducer and is received by a receiving transducer. The pulse is amplified and its amplitude is displayed for analysis on a cathode-ray oscilloscope. If the wave of acoustic energy encounters a discontinuity in the test material (such as a crack or a void), it will be diffracted around the discontinuity thus increasing the distance and time of its travel from the transmitting transducer to the receiving transducer (which, in turn, decreases its amplitude). Therefore, all other conditions being constant, the travel time of the wave in sound material will be less than that for a material containing discontinuities.

There are three commonly used methods of introducing ultrasound into a specimen:

1. *Straight beam* (uses longitudinal waves). The transducer is placed directly on a specimen and sends a straight, perpendicular beam of acoustical energy into it. Signals appear on the cathode ray tube (CRT) at the receiving transducer (on the far end) which can be read and interpreted by an experienced evaluator. Any flaws large enough to intercept the sound beam will show as abnormal patterns on the CRT. Detection of defects may be either by the pulse-echo method or by the through-transmission method. In the pulse-echo method, a transducer emits short pulses of sound and receives the reflections of these pulses resulting from discontinuities in the intervals between pulse emissions. For the through-transmission method, two transducers (a transmitter and a receiver) are positioned on opposite sides of the specimen. Any significant reduction in amplitude of the signal received is an indication of the presence of flaws. If no signal is received, this is an indication that the flaw is large enough to block the sound beam completely. This two-transducer method is used where the shape or internal condition of the specimen does not allow the reflection of the beam back to the same transmitting transducer.

2. *Angle beam* (uses transverse waves). The transducer is mounted on an angle which causes the sound to bounce through the specimen until it strikes a flaw and reflects back to the transducer. This is used on specimens where mounting the transducer directly above the flaw is impractical. The CRT

display is similar to the display produced by the straight-beam method mentioned above.

3. *Surface wave* (uses Rayleigh or Lamb waves). When the transducer is mounted so that the angle of the sound emission is about 63° to the surface of the test material, all of the ultrasonic energy that enters the specimen does so as surface waves (which penetrate only to a depth of about one wavelength). Therefore, this method is useful only for detection of surface or near-surface discontinuities. An advantage to the use of this method is that surface waves will follow the contour of the test specimen.

Advantages and Limitations. Ultrasonic testing is one of the most popular nondestructive test methods for metal evaluation available today for many reasons, some of them are:

1. High sensitivity permits detection of defects about the size of a wavelength.
2. Ultrasound can penetrate thick materials and measure thickness.
3. It can determine position and approximate size of internal defects with accurate and fast readout of results.
4. It can be used with automated testing equipment.
5. It allows specimen testing when only one side is accessible.
6. There are no major hazards to personnel which require special shielding for safety.
7. It is portable and versatile.
8. It is well suited for detection of laminar-type defects that are oriented perpendicular to the energy beam.

Ultrasonic waves do not travel through air effectively. Therefore, a liquid coupling material (such as oil, grease, or glycerine) must be used between the search unit and the test specimen. In immersion testing, a few inches of water is used between the search unit and the specimen.

Certain test conditions may limit the effectiveness (or prohibit the use) of ultrasonic testing. The most common of these are unsuitable geometry and unsuitable internal structure of the test specimen. Specialized training of an operator to interpret the results of this test is required, which makes the test very operator dependent.

Standard calibrated reference blocks must be used to determine operating characteristics of the instruments and transducers to established reproducible test conditions.

Radiographic Evaluation (X-Ray)

Internal Structure and Thickness of Metal

X-ray evaluation techniques are based on the principle that x-ray attenuation is affected by the density, chemical composition, and thickness of the test specimen. Equipment of various types is available for x-ray evaluation. The main ones use photographic emulsion, fluorescent screen, semiconductor conduction values, ionization of a gas or an electron multiplier (or combinations thereof) as the means of detection of the test parameters. Selection of the particular equipment to be used is affected by the following factors:

1. Density and thickness of the test specimen
2. Number of radiographs required per unit of time
3. Geometry of the material
4. Whether the test specimen can be brought to the test equipment or must be conducted in place.

In nondestructive testing of the quality of metal materials, x-rays are used mainly to: (1) inspect for interior soundness; (2) to locate porosity, slag inclusions, incomplete penetration, cracks, icicles, burn-throughs, and other defects in welds; and (3) to detect defects within castings, pressings, and worked or formed metallic parts. Voids measuring 0.5–1.0% of the thickness of steel or aluminum alloy can be detected when using this technique of nondestructive testing under carefully controlled conditions; however, sensitivities of 2% are relatively easy to obtain and are used as the commonly accepted minimum variation.

Advantages and Limitations. X-ray testing offers the advantage of providing a high-resolution picture on a permanent record (film) or a "live" image on an electro-optical imaging system. It is especially useful for detection of laminar-type defects if they are parallel to the energy beam. Since the equipment is heavy and costly, it is used mainly in laboratory evaluations. X-ray sources are harmful to organic tissue, and the test environment must be shielded (or isolated), as necessary, to prevent radiation hazards. Another drawback is the high initial cost of the equipment and film, the time delay required to develop the film, and the requirement that access to both sides of the test specimen is necessary to conduct the test. The user should be aware that there are many films from which to choose; careful selection must be made of speed and resolution in order to obtain the desired results. The entire process of selecting the equipment, operating the equipment, and interpreting the test results requires a high degree of technical training.

Eddy Current

Eddy current testing is one of the most widely used of nondestructive testing methods. The test is based on the measurement of changes in the impedance of a coil due to changes in the flow of eddy currents induced in a conductor. Test material properties and constituent variations that affect the flow of the induced eddy currents sufficiently can be detected, measured, and analyzed. The basic properties of conductors that influence eddy currents are (1) electrical conductivity, (2) magnetic permeability, (3) mass, and (4) discontinuities.

Since so many factors affect the flow of the induced eddy currents, this test method can be applied to a wide variety of test situations. Typically, it is used for thickness gaging; flaw detection; measuring coating thickness; alloy sorting;

monitoring hardness, grain size and heat treatment; and evaluating the condition of wire, tubing, local regions of sheet metal, and welded pipe.

The basic operation is as follows:

1. A signal (10–250 kHz) is generated by a variable oscillator and is applied through a modified Wheatstone bridge to a coil which is placed on a sound piece of material similar to the one being investigated.
2. The bridge is balanced.
3. When the coil is passed over a discontinuity of the test specimen, a change occurs in the impedance of the coil due to the effect of induced eddy currents. This unbalance is recorded on a microammeter.

Many types of probes are used with eddy-current testing equipment, depending on the application. In some cases multiple coils are even used. For example, to detect surface discontinuities in ferromagnetic materials, two driving coils are arranged around the pickup coil in the probe.

Advantages and Limitations. The advantages of using this test method are: (1) it is very fast; (2) it is relatively inexpensive; (3) it is readily automated; (4) the equipment is easily portable; (5) a permanent record can be obtained of the test results if needed; (6) the equipment is simple to operate; and (7) the method can be adapted to many comparative analyses. Limitations to its use are: (1) it is useful only on electrically conductive or magnetically permeable metals; (2) it has a shallow penetration; (3) the use of reference standards is essential, but unfortunately no generally accepted standards are available for ferromagnetic materials to establish the relationship between physical properties and the eddy-current data; (4) only a qualitative comparison is possible, rather than absolute measurements of the flaws; and (5) because eddy currents are distorted at the edge or end of a test specimen, it normally is not practical to inspect flaws any closer to the edge than ⅛ in. Practical experiments have shown that a fairly good correlation between eddy-current data and actual crack depth in ferromagnetic materials is available only up to approximately 0.6 mm (0.025 in.) maximum depth. The test normally is not used to detect discontinuities deeper than 6 mm (0.25 in.).

Magnetic Particle

Surface and Subsurface Discontinuities

The magnetic particle test is a convenient in-situ method of detecting surface and subsurface discontinuities in metals. An electrical current is used to induce a magnetic flux into a specimen, which then produces a patterned magnetic field in the specimen. An indicator made of colored magnetic particles is applied to the test specimen; the particles form visible patterns around the discontinuities according to the magnetic leakage fields formed by the discontinuities.

The surface indicator buildups which are caused by surface discontinuities are thin and sharp, while buildups near subsurface discontinuities appear broad and less well defined.

Various types of electrical currents are used to produce the magnetic fields. The accompanying table summarizes them and describes their uses, advantages, and disadvantages.

Types of Magnetizing Currents Used in Magnetic Particle Testing.

CURRENT	USE	ADVANTAGES	DISADVANTAGES
Alternating current (ac)	For detecting surface discontinuities	Gives best sensitivity for detecting surface discontinuities; relatively easy to demagnetize; no penetration of flux into specimen; best for detecting fatigue cracks in ferrous structural members. Particle mobility is good	Ineffective for detection of subsurface discontinuities
Direct current (dc)	For detection of both surface and subsurface discontinuities; best used mainly for subsurface	Penetration of flux into specimen permits detection of subsurface discontinuities	Must use a fixed voltage. Battery maintenance is required. Difficult to demagnetize
Half-wave rectification (HWAC)	For detection of surface and subsurface discontinuities; most sensitive for subsurface discontinuities	High flux density. Full penetration into specimen permits detection of subsurface discontinuities. Some ac equipment can be converted by adding a rectifier and switch. Particle mobility is good	

There are many means of inducing a magnetic flux into a specimen and producing a magnetic field. Suffice it to say here that the means should be selected to meet best the application and the nature of the discontinuity to be detected.

The indicator particles are made of carefully selected magnetic materials of the proper size, shape, permeability, and retentivity. Two classes of magnetic particles are available for use with this inspection method—wet method particles and dry method particles. Wet method particles use a liquid vehicle, and dry method particles are borne by air. Particles used in the wet method are suspended in oil or water and are available as a powder or thick paste colored either black, red, or fluorescent. Fluorescent particles must be viewed under a "black light" in a darkened room. Particles used in the dry method are in the form of a dry powder in red, black, or gray color. In all cases, color is chosen to give the best contrast on the metal specimen being inspected. The powder is sprayed gently and evenly onto the metal surface where it is free to be influenced by the magnetic leakage field and to form indications of the discontinuities.

The advantages to using this test method are:

1. It is relatively inexpensive.
2. It provides instant results.

3. It is portable.
4. It is easy to use with little training.
5. It is sensitive to surface cracks.
6. It is a nondestructive test method.

Limitations to its use include:

1. It requires external power supply.
2. It is limited to use on ferrous materials.
3. The evaluator must have a thorough knowledge of the test and be able to evaluate and record test results with accuracy.
4. Sharp, angular types of discontinuities are more easily found than round or streamline types.
5. Surface discontinuities are more easily found than those located below the surface.
6. Retained magnetic fields hinder cleaning of the specimen.
7. The specimen may require demagnetizing after the test.
8. Wet particles are best suited for detection of fine surface cracks (such as fatigue cracks).
9. Dry particles are most sensitive for detecting subsurface defects and usually are used with portable equipment.
10. Some surface preparation is needed.
11. Coatings can affect the ability to magnetize the specimen.

Destructive Evaluation of Metals

When evaluating metal structures to determine their ability to perform well in a rehabilitation situation, the most important material properties to determine are:

1. Yield strength
2. Yield point
3. Tensile strength
4. Elongation (ductility)
5. Modulus of elasticity
6. Compressive strength.

Using destructive methods of tension or compression testing of samples, known as coupons, the properties are determined as follows:

Tensile Testing

Yield strength
Yield point
Tensile strength
Elongation (ductility)

Compression Testing

Yield strength
Yield point
Modulus of elasticity
Compressive strength (for *some* materials)

Coupon Test

Strength, Ductility, Modulus of Elasticity

In the coupon test, a test sample (known as the coupon) is removed from the metal being evaluated and is tested for the important material properties indicated above. The test machine must conform to the requirements of ASTM (Methods E 4), Verification of Testing Machines.

Test procedures for tension or compression testing of coupons are described in "Standard Methods of Tension Testing of Metallic Materials," ASTM E 8, and "Standard Methods of Compression Testing of Metallic Materials at Room Temperature," ASTM E-9 (American Society for Testing and Materials, 1977). The tests are applicable for use on (1) die castings, (2) forgings, (3) structural shapes, (4) malleable iron, and (5) powdered metals. Coupon shapes can include:

1. Plate-type specimens
2. Sheet-type specimens (sheet, plate, flat wire, strip, band, hoop, rectangular, etc.)
3. Round specimens
4. Rectangular bars
5. Pipe and tube shapes.

The test coupons can be substantially full size or machined, as prescribed in the product specifications for the material being tested. The shape of the ends of the specimen outside the gage length must be suitable to the material and must be of a shape to fit the holders or grips of the tension testing machine so that loads are applied axially. Figures 1–5 of ASTM E 8 describe gripping and snubbing devices used with tension testing equipment, while Figures 6–9 describe the shapes and dimensions to be used for tension coupons. Similarly, a description of coupons and jigs used for compression testing is provided in Tables 1 and 2 and Figures 1–4 of ASTM E 9.

Location of Coupons. Coupons are taken from locations in the metal material to be tested; and, generally, the axis of the coupon is located as follows:

1. At the center for products 38 mm (1½ in.) or less in thickness, diameter, or distance between flats
2. Midway from the center to the surface for products over 38 mm (1½ in.) in thickness, diameter, or distance between flats
3. For forgings, specimens are taken either from the predominant or thickest part of the forging, or coupons may be forged separately as representative samples of the forging being evaluated. Unless otherwise specified in the applicable product specifications, the axis of the specimen should be parallel to the direction of grain flow.

Surface Finish for Coupons. Particular attention has to be given to the uniformity and quality of surface finishes of specimens for high-strength and very low-ductility materials, since they can cause variation in test results. Surface finish of the coupons is described in applicable product specifications.

Procedures for Performing Test. Procedures for preparing and measuring test specimens and for conducting tension or compression tests are described in paragraph 5 of ASTM E 8 and paragraphs 5–7 of ASTM E 9, respectively, and include a description of the speed of testing (rate of separation or closure of heads, total elapsed time, rate of stressing and straining, etc.). Stress-strain diagrams for determining yield strength and yield strength and yield point are provided in Figures 20–22 of ASTM E 8, and Figure 5 of ASTM E 9 for tension coupons or compression coupons, respectively.

VII. Concrete

CONCRETE—THE GROUND RULES

A suitable quality of concrete means simply concrete that will perform satisfactory service in the use for which it is intended. To do so it must possess four essential functional properties: strength to carry superimposed loads; watertightness to prevent water penetration; durability for wear and weather resistance; and workability to ensure proper handling, placing, finishing, and curing. These properties can all be achieved with good materials and careful workmanship, but it is not simply done nor easy to do.

Sound materials must be selected, carefully proportioned, and combined. A low water-cement ratio must be maintained and the temptation to add water to place the concrete more easily must be controlled. The right mixture of workable consistency to create the required surface finish must be maintained. When this is done the concrete must be adequately cured at the right temperature and humidity. All of this must be carried out by directing dozens of men, many of whom do not take kindly to direction.

The correct mixture of materials will determine the desirable qualities of the

Precasting concrete panels, New Jersey. (Photo by F. Wilson)

Precasting concrete panels, New Jersey. (Photo by F. Wilson)

Poured in Place platform facia, lower Manhattan. (Photo by F. Wilson)

plastic concrete, which are consistency, workability, and finishability, and ensure the essential properties of the hardened concrete which are strength, durability, watertightness, wear-resistance, and economy.

In properly made concrete, each particle of aggregate, no matter how large or small, is surrounded completely by a cement-water paste, and all spaces between aggregate particles are filled completely. The aggregates are considered to be inert materials bound together by the cementing medium into a solid mass. The cementing property of the paste develops in a chemical reaction between the cement and water.

The quality of the concrete is dependent on the quality of the paste; it is essential that the paste have strength, durability, and watertightness. The critical water-cement ratio is expressed in weight of water to weight of cement; this proportion was formerly stated in gallons of water per 94 pound sack of cement.

Only a relatively small amount of water, about 3½ gallons to a sack of cement or a ratio of 0.31, is necessary to complete the chemical reactions between water and cement. But more water, ranging from 4 to 9 gallons to the sack of cement or a water-cement ratio ranging from 0.35 to 0.80, is used to make the mixture workable. With more water more aggregate can be used, with resulting economy of material, for the cement is the most expensive part of the mixture.

There is a direct link between strength of concrete and the water-cement ratio, providing, of course, that the mixture is plastic and workable. For given materials and handling conditions, strength is determined primarily by this ratio.

Habitat, Montreal, under construction. (Photo by F. Wilson)

Precast concrete building system, Moscow. (Photo by F. Wilson)

More water will result in less strength, less water in greater strength. For a given water-cement ratio in a concrete mixture, the strength at a certain age is predictable, assuming that the mixture is plastic and workable, aggregates are strong, clean and, sound, and the proper curing care is taken.

Too much water results in a diluted paste and a weak and porous concrete when it hardens. Not enough results in a mix that cannot be properly placed and finished. Cement paste made with the correct amount of water has strong binding qualities, is water-tight and durable. If the cement paste and the aggregates are strong and durable, the concrete is strong and durable. If the cement paste is watertight, the concrete is watertight. Strength, durability, freeze-thaw resistance, watertightness, and wear-resistance of the paste, and therefore the concrete, are largely controlled by a sufficiently low ratio of water to cement. The most destructive of the natural forces of weathering is freezing and thawing action on wet or moist concrete.

It would seem that such simple relationships between ingredients and clearly defined procedures would make concrete an ideal building material. Somehow, although concrete has proven a strong, adaptable building material it has proven far from predictable. At the present time a great deal of difficulty is being experienced with concrete. The diagnosis of concrete to correct past difficulties, maintain present quality, and ensure as few aging difficulties as possible for the future, requires the combined skills of historians, engineers, architects, and builders. Some notes on the delicate art of balancing the simple relationships of this material follow.

Capsule History of Reinforced Concrete

Problems with concrete are not new; they have only become more visible. Concrete has been widely used for the past 50 years. We find it in reinforced structural frames, cast stone, concrete masonry units, and as both precast curtain wall and precast structural elements.

Concrete, according to Prouden (*P/A*, 11/81) has reached a critical stage in its aging process. An entire generation of reinforced concrete and cast stone buildings built between 1900 and 1940 are beginning to demonstrate signs of failure, requiring extensive repairs.

The recent history of concrete construction may reveal some of the causes of this phenomenon. The quality of cement, concrete, changes in the cement industry and building practice have all contributed to our present problems. The history of concrete construction is closely tied to the history of the development of the iron and steel industries as well.

Beginning at the end of the 18th Century and well into the 19th Century Britain was bombarded by extravagant claims for products advertised as artificial cements. Although these products were constantly improving, their variety, according to Prouden, defies description. They included, among others, hydraulic limes, Roman cements, and Portland cements. Aspidin's patent in 1824, although a landmark, was only one in the steady development of cementitious materials throughout the 19th Century.

An important development was the discovery and improvement of natural cements during this time. John Smeaton's work on the Eddystone Lighthouse and the later "Roman cement" patented by James Parker are examples of natural cements with hydraulic characteristics. In North America, Canvass White discovered natural cement rock in 1818 during the excavation of the Erie Canal. He started a cement works near Rosedale, New York, and by 1839 the name of Rosedale became synonymous with natural cement. The product was used on such famous structures as the Brooklyn Bridge, the Brooklyn Navy Yard, and the California State Capitol.

Natural cement is inconsistent in quality, for it depends to a large extent on the composition of natural cement rock. Therefore the work of Joseph Aspidin and others in England, which led to the creation of Portland cement with its controlled qualities, was important for building. Portland cement was being manufactured by David Saylor in Pennsylvania by 1871, but until the 1900s most Portland cement used in America was imported.

The use of concrete increased rapidly during the last half of the 19th Century and first quarter of the 20th. Major developments took place in the manufacture of cast stone and in the use and application of steel reinforcement in monolithic construction.

The use of cast stone was related to the increasing cost of natural stone and its carving for building ornament. The terms *artificial stone, manufactured stone,* and *architectural stone* came into use at this time to describe these small prefab-ricated units. Various techniques to simulate stone in precast concrete, rustica-tion, veining, tooling and ornamental detailing were common well into the 20th Century.

The history and development of reinforced concrete construction is tied to the evolution of iron bars, beams, and rods, in combination with masonry or concrete used for fireproof floor systems. A fireproof concrete floor system with steel cables in the concrete was patented by William B. Wilkinson as early as 1854. Other patents were also developed; one used wire mesh in reinforcing concrete garden furniture in 1867.

Earnest Ransome began to develop reinforced concrete buildings in the 1880s. The widespread interest in the potential of the material involved inventors like Thomas Edison, who manufactured Portland cement and built several concrete buildings. He held several patents for poured-in-place formworks for use in low-income housing.

Reinforcing iron and steel bars in concrete "rebars" originated about 1878, according to Dorsey, and specifications for them were first developed by the Association of American Steel Manufacturers in 1910. The American Society for Testing and Materials adopted standard specifications for billet steel concrete reinforcing bars in 1911.

The American Concrete Institute was organized in 1905 and the Portland Cement Association in 1916. Both helped to establish and maintain standards and quality control procedures.

With the development of material sciences and engineering, confidence in reinforced concrete began to grow. Reinforced concrete structural frames became common, although clad with masonry materials or cast stone. As the use of concrete developed, concrete became an acceptable architectural material in its own right; exposed concrete frames became commonplace. The work of Frank Lloyd Wright helped explore the aesthetic potential of concrete. The Second World War gave further impetus to concrete construction due to the shortage of steel.

Despite spectacular advances in cement and steel, concrete remains an exper-imental material. The reason for its difficulties and successes are not always consistently explained. A great deal of analysis and evaluation of its performance has been done. What follows here is what is known today. It is bound to change, for the use of post-tensioned, prestressed concrete and new additives such as plasticizers have withstood the test of theory and laboratory in full measure but have not yet stood the test of time.

The Consequences of Placing Steel in Concrete

An elementary understanding of the interaction between steel and concrete is important if the forces responsible for concrete cracking are to be diagnosed.

A steel reinforcing bar embedded in concrete is stressed as the concrete dries and shrinks. Since the steel restrains the concrete from shortening it is subjected to tensile strain and tensile stress. When the stress reaches the ultimate strength of the concrete, at that age, the concrete cracks. There is a race between the increase in strength of the concrete over time and the increase in shrinkage strain as it dries. The advantages of prolonged curing periods, during which the concrete can gather strength as it slowly shrinks, are self-evident.

When the concrete cracks, strain on the concrete is relieved, as is the stress at that point. With nonreinforced concrete the crack opens as the concrete shrinks. The crack reduces the stress to zero. Steel bars change this situation. When the concrete cracks the steel attempts to hold the two sides together and is in tension. The steel, at this point is subject to deformation equal to the width of the crack. The stress on the steel would be extreme, and if the width of the crack indicated the strain on the steel it would fail. But deformation is distributed,

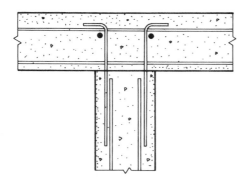

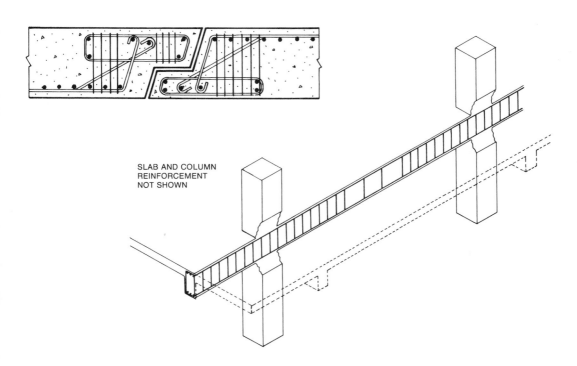

SLAB AND COLUMN
REINFORCEMENT
NOT SHOWN

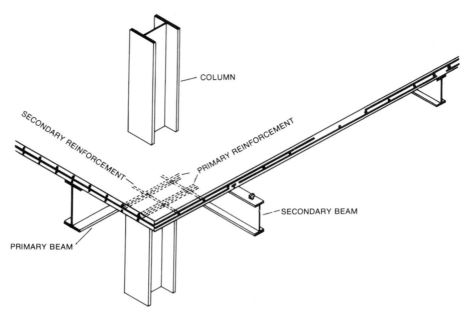

COLUMN

SECONDARY REINFORCEMENT

PRIMARY REINFORCEMENT

SECONDARY BEAM

PRIMARY BEAM

(From Structural Systems, Cowan and Wilson, Van Nostrand Reinhold, New York, 1981)

as a strain, along the bar for some length. This is provided by local bond slippage or creep between the steel and the concrete on each side of the crack.

Away from the crack and beyond the bond slippage concrete strain will be somewhat relaxed, but total relaxation is resisted by the steel in compression. The tensile force in the concrete is balanced by both the compressive force in the steel and the tensile force in the steel at the crack, both acting in the same direction.

The compressive stress in the steel is due to concrete shrinkage and is quite small. It is therefore neglected in calculations. The tensile force in the concrete is equal to the tensile force in the steel.

As a rule of thumb calculating the tensile strength of concrete is considered to be $\frac{1}{10}$ its compressive strength. For a 4,000-psi concrete a figure of 400 psi is used for tension. But 4,000 psi is the concrete's strength at 28 days. Cracks develop before the concrete comes to its 28-day strength. Some studies indicate that a prime cause of cracking in concrete is the thermal contraction of cooling following the heat of hydration. In the relatively thin sections used in buildings heat is largely dissipated in about 3 days. It is therefore suggested that the tensile strength of concrete at 3 days should be used. This can be assumed to be about $\frac{1}{3}$ or $\frac{1}{2}$ the 28-day tensile strength, or $\frac{1}{30}$ or $\frac{1}{20}$ its crushing strength.

The spacing of cracks is an important indicator of the stress and strain of the member. Concrete stress is reduced to zero at the crack and gradually increases along the length of steel where bond slippage has occurred. At the end of this length, cracking stress conditions are reintroduced and a second crack may develop.

SOURCE

Cracks, Movements and Joints in Buildings: Recordings No. 2 NRCC 15477 Division of Building Research, National Research Council Canada, 1976, Record of the DBR Building Science Seminar, Autumn 1972.

All Inanimate Objects Tend to Disintegrate

All inanimate objects tend to disintegrate, and concrete finds more exotic ways of doing so than by simply cracking around steel reinforcement. The inclusion of reactive ingredients or embedded materials in the concrete mix helps speed this natural action. The proper concrete mix becomes chemically inert after the cement reactions are complete, but certain chemicals can interfere with this action. They react badly with the cement paste that holds the concrete together. These chemicals include the entire family of sugars, lignites, resins, sulfur vapor, and salts, to name a few among many others.

Reactive rocks used as aggregate are a well known cause of strange concrete behavior, but such knowledge is sometimes not heeded or ignored. A spectacular example of this odd behavior was exhibited in the use of dolomite as a coarse aggregate in the construction of two 640-ft-long concrete dry docks during WW II. They have been expanding ever since. The docks are now 20 in. longer than they were when originally constructed, and a good deal of this increase in size is in the form of cracks.

It has been well known that the use of sugar will inhibit the setting of concrete. On a New York grade crossing, when the concrete forms were stripped a soft spongy brown area was exposed. It seems that the workmen had jettisoned coffee containers into the forms. The entire area had to be dug out and unsugared concrete substituted.

A great number and variety of aggregates have been tried in concrete. For a time one of the most popular was sawdust. Concrete with sawdust aggregate reached a strength of 2,500 psi, was light and nailable but the practice was discontinued when a theater lobby floor caught fire in the New England area.

Cinder concrete was and remains a popular lightweight aggregate, but it is only satisfactory when good, clean, completely combusted anthracite clinkers are used. Bituminous cinders that have not been completely burned deposit free sulfur in the slab, which will corrode the reinforcing causing cracks and spalls. If the building is near the seashore the condition is aggravated by the moist, salt-laden environment.

A most curious difficulty was encountered in the perforation of concrete casings around wooden piles in 1923. Investigation revealed that the holes found in the concrete were the result of the concrete's infestation by small borers. These were native to the local rock used in the concrete mix. The deterioration, however, was not actually due to the borers themselves, but the chemical action from the wastes of the growing mollusks.

Although concrete might indeed become as hard and impervious as rock, it still suffers from the things builders insist upon inserting during its plastic state.

For example steel conduit and reinforcement embedded in concrete with a residue of sodium chloride and other salts will rust if moisture penetrates the slab. Electrical conduit can cause cracking and spalling problems in concrete ceilings if the metals in the slab are not metallurgically identical: if the conduit does not match the metal alloy of the boxes, or if the conduit is not of the same metal as the reinforcing steel, difficulties may well ensue. This was proven disastrously in New York City when aluminum conduit was permitted by building regulations for a short period of time.

These are but a few of the stories of mishaps. This study is not addressed particularly to the description of such exotic malfunctions. It instead concentrates on the common, less dramatic occurrences that plague designers, engineers and builders. The foregoing descriptions were to remind the reader that although the Pantheon and Hadrian's tomb were built almost two thousand years ago and still stand quite satisfactorily, the practice of concrete construction is not quite yet an exact art.

REFERENCE

Feld, Jacob, *Failure Lessons in Concrete Construction,* A Collection of Articles from *Concrete Construction* Magazine, Addison, Illinois, 1978.

CRACKS IN CONCRETE—CAUSES AND CURES

Cracks seem inevitable in concrete. Whether they are or not is another question. Professor Joe W. Kelly states that they rarely affect the structural action or the durability of concrete significantly, but they are not reassuring in appearance. They also let water into the building more easily and may accelerate weathering or rusting in some instances. On the whole almost everyone would be happier without them.

The technique of making crack-free concrete is not yet known according to Kelly, but we do have enough knowledge to prevent many cracks and avoid many more. However, there is no simple or sure formula for doing so because there are so many causes. Not all causes are subject to control simultaneously with the great number of people involved.

Kinds of Cracks

Cracks come in all sizes, shapes and configurations. If we categorize by depth there are surface, shallow, deep, and through cracks. If we label them by surface direction there are essentially two varieties. The first is "map cracks" or "pattern cracks." These are uniformly distributed short cracks running in all directions, forming rough hexagonal patterns. Map and pattern cracks are a signal that the surface layer of the concrete has been restrained by the inner or backing concrete. The second surface category is the single, continuous crack. These know where they are going and run in definite directions, often parallel and at definite intervals. They indicate restraint in the direction perpendicular to them.

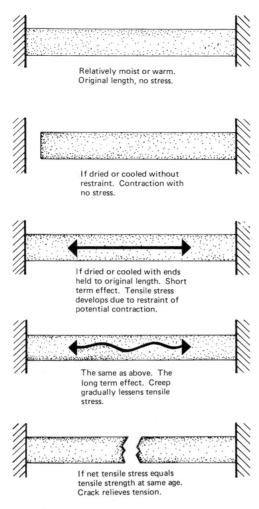

Relatively moist or warm. Original length, no stress.

If dried or cooled without restraint. Contraction with no stress.

If dried or cooled with ends held to original length. Short term effect. Tensile stress develops due to restraint of potential contraction.

The same as above. The long term effect. Creep gradually lessens tensile stress.

If net tensile stress equals tensile strength at same age. Crack relieves tension.

(After Joe W. Kelly, *Concrete Construction Magazine*)

There are also internal cracks around larger pieces of aggregate, cracks in fresh concrete, and cracks that occur after the concrete has hardened. There are incipient cracks in fresh or "green" concrete which may only appear later if stressed by other forces. There are unsoundness cracks caused by chemical reaction, and a great many more, says Professor Kelly. What follows deals mostly with continuous, directional, "everyday" cracks in hardened concrete, although it might apply to some of the other categories described.

The Cracking Process

A simple statement concerning a very complex series of happenings is that concrete cracks when the stresses exceed its strength. The circumstances that may lead to such failure are many indeed. Given a concrete bar that dries or cools without restraint we will find that it simply contracts and no stress is developed. If, however, it is restrained during drying or cooling and the bar is forced to maintain its original length, then tensile forces will be induced in it. If it is strong enough to withstand them for a period of time they will eventually be relieved somewhat by creep, or plastic flow. At any time within this process whether the concrete is fresh or hardened, if the net tensile stress builds up to the tensile strength at that age of the concrete, the concrete will crack and the stress will be relieved. This is a simple example but it illustrates the process which takes place at larger or smaller scale in all kinds of locations due to many causes.

It is evident that no one property can be singled out as the sole or the principal cause of cracking. Boyd Mercer of Australia classified the causes in a table.

Causes of Cracking of Concrete

Before hardening
  Constructional movement (subgrade, formwork)
  Settlement shrinkage (around reinforcement, obstructions, aggregate)
  Setting shrinkage (plastic, early-age)

After hardening
  Chemical (cement constituents, carbonation, reactive aggregates, foreign bodies, rust)
  Physical (drying shrinkage, moisture fluctuations)
  Thermal (internal heat-of-hydration temperature stresses, differences in thermal properties of aggregates, external temperature variations, frost and ice action)
  Stress concentrations (reinforcement, structural form as at corners of openings, creep)
  Structural design (loads, foundation settlement, etc.)
  Accidents (overload, vibration, fatigue, earthquake, fire)

Factors Affecting Cracking

Restraint is the most obvious cause of cracks. When concrete wishes to stretch or shrink, or alter its dimensions and is restrained, it cracks. A concrete slab which butts against a building or other mass of dissimilar size must be isolated along all points of abutment to accommodate differential movement.

Shrinkage is the reduction in volume caused by loss of water. Loss of water takes place both physically and chemically. Physical loss of water is largely by sedimentation, which is called ''bleeding,'' and by subsequent evaporation. As concrete experiences drying, shrinkage forces within the slab must be controlled by contraction joints every 20 feet. If the mix was soupy, that is, contained excessive water, shrinkage will be more pronounced.

Reinforcement for structural stresses should be set at from ⅓ to ¼ from the bottom of the slab, and reinforcement to control shrinkage should be set from ⅓ to ¼ from the top of the slab, top and bottom steel. The size of reinforcement needed for structural stress is considerably larger than that required to control surface cracking.

Another kind of cracking, *plastic cracking,* may occur. This is caused by a change in weather conditions which brings about an increase in the rate at which water is evaporated from a fresh concrete surface. When wind velocity, relative humidity, high air temperature, or all three combined cause water to evaporate from the concrete faster than it is replaced by bleeding, cracks will result.

Some of these factors and others affecting cracking, as best we can tell given the limited information available, are detailed below.

Water. The amount of water per sack of cement or per cubic yard of concrete is an important factor, probably the most important. For the more water the greater the tendency to crack, since water both increases shrinkage and reduces strength.

Cement. The amount of cement is important, for the richer concretes crack more. Finely ground cements and cements high in silica rate relatively high in shrinkage but not necessarily high in cracking. Carbonation of cement from the carbon dioxide in the air produces initial contraction and reduces subsequent drying shrinkage.

Aggregate. The mineral composition, shape, surface texture, and grading of aggregates affect the proportions, thermal coefficient, drying shrinkage, stiffness, creep, and strength of concrete. It is a good practice to look at structures in use to compare cracking caused by aggregates, although cracking can also be influenced by the conditions of mix, weather, and placing. It is quite well known that certain clays in aggregates cause high shrinkage and cracking, since clay shrinks more than cement gel. It should also be taken into consideration that the smaller the maximum size of aggregate the greater the shrinkage will be of the concrete. Larger rock aggregates restrain shrinkage locally more than smaller pieces.

Admixtures. These may affect cracking because they influence the rate of hardening, shrinkage, and creep.

Bleeding. The upward flow of water in fresh concrete causes pockets of watery paste under the larger pieces of aggregate, especially in deeper sections, and breeds internal cracks.

Placing. The rate and conditions of placing undoubtedly affect cracking through such influences as bleeding, segregation in forms and around reinforcement, and temperature. Uneven settlement of slab subgrades causes many cracks.

Curing. Moisture conditions during curing seem to be highly important. The rapid drying of fresh concrete in slabs within minutes of placing may cause evaporation exceeding the rate of water bleeding to the surface. When this happens, according to William Lerch (*ACI Journal,* Feb. 1957) "the surface of the concrete has attained some initial rigidity; it cannot accommodate the rapid volume change of plastic shrinkage by plastic flow (creep); and it has not attained sufficient strength to withstand tensile stress. Thus plastic shrinkage cracks may develop."

After the process of hardening begins, one single physical agency, evaporation, becomes important. The shrinkage forces then act relative to the two main

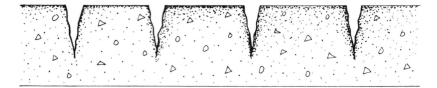

Cracks may be the result of rapid drying of the upper surface of the concrete. It shrinks rapidly stressing itself before it has gained enough strength to withstand the strain of the pull against the interior mass of concrete.

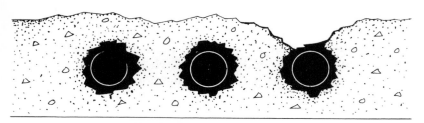

Cracks and destruction of the concrete surface may be caused by the expansion of reinforcement due to oxidation.

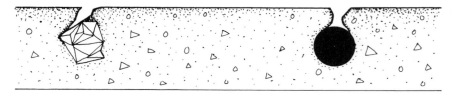

Concrete may shrink around solid objects below its surface such as large pieces of aggregate, rocks left in the forms or reinforcement too near the surface, causing cracks.

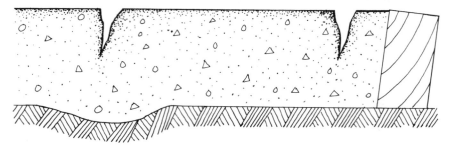

Concrete cracking may be caused by a shifting of its supporting surface or form deflection before it has developed enough strength to withstand the stresses caused by these movements.

components of the mass, the cement paste and the aggregates; the interaction between these two can lead to cracking.

Evaporation is also the main cause of the plastic shrinkage cracks which result from uneven, accelerated loss of water. The effect of evaporation is influenced to a large extent by such climatic conditions as intensity of air movement and temperature and humidity changes.

In hydration, the loss of water by the chemical reaction of the hardening process, the water present in the mix becomes a component of the solid products of the process without any loss of weight occurring. The reaction is accompanied by a change of volume; the water loses about 25% of its volume in combining with the compounds present in the cement. In wet mixes chemical shrinkage is generally far less important than physical shrinkage, whereas in dry mixes chemical shrinkage becomes increasingly important, even to the point where it may be necessary not only to prevent evaporation of water but also actually to replace the water used up chemically.

Temperature. Temperature influences the rate of strength growth in concrete, but its principal influence on cracking lies in its fixing the "base" length during the early hours when the concrete becomes rigid. A subsequent lowering of

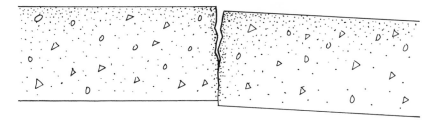

Cracks may be caused by a heavy load imposed on concrete with insufficient reinforcing to counteract the strain. They may also be caused by a shifting of the subgrade downward. The concrete cannot sustain its own weight without this support. Solid may also expand upward cracking the concrete.

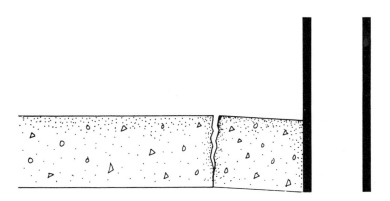

Cracks may be caused by uneven settlement of building elements such as a cellar slab on grade connected to a foundation wall or metal column. If the two elements are not separated so they can act independently cracking will occur.

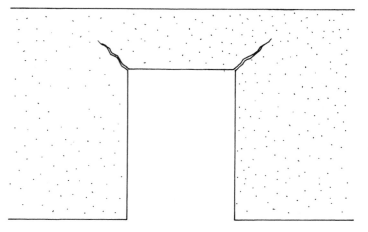

Cracks may appear at sharp corners due to the concentration of forces in these locations.

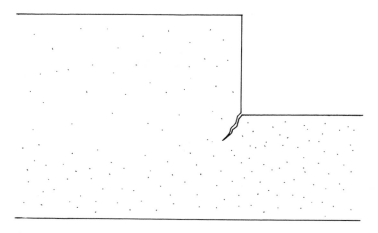

Cracks may appear at the corners of abrupt changes in building sections.

temperature will produce potential contraction. This effect is especially important in hot weather and in mass concrete. Sections of pavement cast during cool weather are less subject to cracking than those on the same job cast during hot weather. This is true to various extents for other structures. It might be better to pour concrete during the lower temperatures at night to minimize some cracking.

Exposure. Weather conditions of exposure influence cracking to a great extent. Steep thermal gradients and moisture gradients produce severe internal restraints between surface and interior or backing. Southern and western exposures are usually more severe than northern and eastern.

Restraint. Over-rigid restraint by foundations or adjacent structural members can possibly lead to cracking. Foundation restraint obviously cannot be prevented

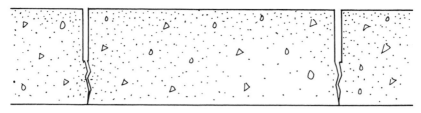

Cracks may even appear where they are meant to be. . . .

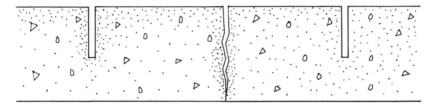

but usually do not.

entirely, and it is a common source of vertical cracks near the bottom of building walls, even if the cracks do not extend to the top. A long wall or slab without joints or breaks is certain to crack at intervals.

As a general rule, the greater the restraint of contraction the more numerous, but narrower, will be the cracks. Heavily reinforcing a wall or slab will result in more but narrower cracks than the conventional light "temperature" reinforcement; however, the total cumulative width of cracks will be about the same. Similarly, high-yield-point reinforcing steel distributes cracks more favorably than structural-grade steel. Narrow cracks are usually preferable, as they are less noticeable and less susceptible to moisture penetration.

A monolithic wall cast integrally with a structural-steel frame would be expected to crack both vertically and horizontally.

Restraint is also induced whenever a portion of a slab or a structure settles or moves differently from the rest.

Internal restraints are caused by differences in richness or composition of mix, moisture content, and temperature between portions of the mass.

It is usually a combination of factors that result in unsatisfactory concrete work. Satisfactory structures have been built and are usually built even with a few unfavorable factors. Concrete, Kelly says, is like a pet dog that will take a lot of abuse and still serve faithfully. If concrete did indeed fail whenever neglected or abused it would probably be treated with more care and respect.

One means of preventing cracking is to use prestressed concrete. Cracking is primarily a tensile phenomenon and prestressing places the concrete in compression. If there is any contraction due to the lowering of moisture or temperature or any elastic strain due to service loads they merely relieve part of the compression.

Cracking Under Load

Cracking under load of a reinforced concrete member occurs under two conditions: when the member is loaded in axial tension and the tensile strength of the concrete is exceeded. Primary cracks passing through the member occur at the weakest sections, which are randomly spaced when the member is loaded in bending and the modulus of rupture is exceeded. Secondary cracks when the loading is increased may also appear. These cracks will run perpendicular to the span of the beam.

If the load is increased still further failure will eventually occur by the crushing of the concrete at the compression face. This failure is preceded by the development of one or more horizontal cracks, and the length of these cracks increases rapidly with increasing load. Such compression failure, resulting in final collapse, usually occurs on a plane at about 30° to the horizontal and located at one or both ends of a horizontal crack.

Failure does not always take place in this manner, as the description presumes a condition of pure bending alone. The actual pattern of failure will depend on the type of loading. When shear loads are also involved, diagonal tension cracks will also appear at, or close to, existing flexural cracks. The shape of these diagonal cracks depends mainly on the length of the shear span; that is, the distance from the support to the nearest load point. As the span becomes longer, the crack will gradually bend over until it runs approximately parallel to the axis of the member. Failure occurs either by crushing of the concrete above the diagonal crack, or by splitting along the level of the main reinforcement. If the member is also loaded transversely, failure is exaggerated by extension of the existing cracks.

The pattern of cracking, its location, depth and width of the cracks, the presence of foreign material on the cracked surfaces, and differences in elevation between two contiguous cracked concrete masses are factors that help determine the cause of the cracks.

Cracks radiating from a floor around a column often indicate that an isolation joint should have been provided.

Cracks on a slab-on-grade floor where one part of the concrete is lower than the other usually indicate differential settlement of the subgrade.

Localized cracking on flatwork in traffic areas is evidence of a slab that is too thin for conditions, has poor positioning of the reinforcement, and/or an unstable sub-base.

In most cases, cracks must be considered active if their cause cannot be determined. Cracks that appear and continue development after the concrete has hardened are also considered active.

Cracking is termed *dormant* when it is caused by a factor that is not expected to occur again. This category includes plastic cracks; cracks resulting from temporary overloading, as from the movement of a piece of machinery over a slab; and random cracks caused by improper timing of a concrete sawing operation. Usually a dormant crack can be permanently repaired after the full extent of cracking has occurred.

Crazing cracks are relatively shallow and form a hexagonal pattern. This phenomenon usually occurs while concrete is in its plastic state. Occasionally it occurs shortly after the concrete has hardened. Crazing is often the result of the surface of the concrete shrinking at a more rapid rate than the interior concrete mass.

Crazing can also be caused by an overly rich mix; too high a slump; poor timing in the finishing operation; excessive finishing; temperature stresses during the early stages of concrete; rapid loss of moisture from the mix in hot, dry, or windy weather; and/or an absorptive subgrade. Although crazing cracks are usually dormant, extreme exposure can result in progressive enlargement of the cracks.

Controlled cracking, as in preformed contraction or expansion joints, ordinarily does not require repair. This type of cracking is either left as is or filled with an elastomeric material. It is random, unplanned cracking, in contrast, that requires extensive repair work.

Other Types of Deterioration

Dusting. The surface of concrete becomes soft and rubs off readily under abrasion or traffic, as a fine, powdery material. Dusting is most often encountered on floors where traffic and abrasion are heavy. Sometimes it is encountered on walls, and can be so severe that it prevents the application of paint or coatings. Common causes of concrete dusting include: concrete mixtures that are too wet; premature or excessive working of the surface; organic materials in the aggregates; and inadequate curing. Another cause is carbonation of the surface due to use of unvented heaters during cold weather.

Efflorescence. Salt crystals on the concrete surface are caused by water that migrates from the interior mass of the concrete to the surface, where it evaporates, are deposited as salts dissolved in water that migrates crystals. This is a common phenomenon in concrete, and is not objectionable unless the deposits are excessive and mar the appearance of the surface.

Form Scabbing. When forms are improperly oiled, the bond between the form face and the concrete fails to break. Consequently, concrete sticks to the form during the stripping operation. The resulting surface has uneven, spalled areas.

Honeycombing. Results when the coarse aggregate is placed with an insufficient amount of mortar. It occurs because the concrete mixture is undersanded and placing techniques are poor.

Deterioration of poured in place concrete due to salts on steps and freeze thaw cycle. (Photo by F. Wilson)

Permeability. Permeability results from cracking, voids, insufficient concrete density, insufficient concrete-member depth, general deterioration of the concrete, or exposure to hydrostatic pressures for which the member was not intended. Permeability is a symptom, not a cause.

Popouts. Popouts are spalls caused by the expansion of a particle that is fairly near the concrete surface. These usually occur in horizontal slabs, and the materials that most often cause them are certain shales, cherts, lignites, limestones, and in tropical regions, iron sulfides. Certain absorbent aggregates that expand when exposed to freezing can also cause popouts. In all instances there must be the presence of moisture to cause the unstable material to expand. Unless popouts occur shortly after the concrete has hardened, they may not be noted for a year or more after the concrete has been placed.

Sandstreaks. These are vertical streaks of sand that appear on the surface of the concrete, and are most noticeable when forms are stripped. This phenomenon can occur when a concrete mix with a high water content or a deficiency of the finer sand sizes is placed in formwork that is not watertight. The sand and water may be forced to the surface, leaving a sandy section on the face of the concrete.

Scaling. This is the sloughing off of relatively thin layers of concrete. Scaling can occur over a limited area or it may be a continuing phenomenon that spreads gradually over virtually the entire concrete surface. Scaling can be caused by

Deterioration of stair carriage due to moisture penetration and subsequent corrosion of stair reinforcing. (Photo by F. Wilson)

Disintegration of concrete parge on brick chimney. (Photo by F. Wilson)

severed freeze-thaw conditions, improper use of de-icing salts, repeated wetting and drying of the concrete, poor finishing practices, incorrect application of a dry shake on floors, chemical attack on the concrete, and heat blast and fuel spillage from jet aircraft.

Spalling. Spalling is a loosely used term commonly referring to chunks of concrete that have been broken from the surface by either form removal or mechanical damage, such as impact on floor joints. Spalling is also caused by corrosion of the reinforcing. The depth and extent of spalling are important in determining the type of repair technique and material to use.

Stains and Uneven Coloring. These can be caused by many materials. Concrete surfaces can be stained by oil, lactic acid, certain syrups, and a number of acids, especially inorganic acids. Stains also result when aluminum and iron are used in or near the concrete. Certain types of form oils, as well as the incorrect application of some form oils, can also severely stain concrete.

Conditions within the concrete mixture itself may also cause uneven coloring or staining. This might be due to differences in the colors of different brands of cement, differences in aggregate colors, the effect of admixtures, and the effect of improper finishing and curing practices.

REFERENCE

Kelly, Joe W., *Cracks in Concrete: Causes and Prevention,* A Collection of Articles from *Concrete Construction* Magazine, Addison, Illinois.

INSPECTION OF CONCRETE

Interpreting results of visual inspection requires understanding of the building, its construction patterns of stress, and the processes of decay in general. Visual indications of deterioration are limited. Cracking in the concrete may be the result of corrosion of the reinforcement, deflection, settlement, thermal expansion, contraction, or shrinkage cracking. The location, configuration, and pattern of these cracks will have significant meaning related to the general context of the building.

Spalling is quite commonly caused by corrosion of the reinforcement, but it may also be attributed to aggregate alkalinity. If staining is evident the color and location of the stains is significant. Brown staining is most often the result of reinforcement corrosion. Occasionally it is caused by aggregate but its pattern should indicate which of the two are responsible. Erosion and disintegration of the concrete surface can be the result of salt crystallization and the freeze-thaw cycles of weathering. A chemical reaction resulting in the dissolving of the cement binder due to weathering can leave the aggregate exposed.

If an extensive evaluation of the concrete in-situ is desired, testing to determine the quality of the material and causes of deterioration is required. Such tests can involve sonic tests, pulse velocity testing, and monitoring of cracks if questions concerning their cause exists. A pachometer may be required to locate and map the location of rebars. Some pachometers will give an indication of relative size and condition of reinforcement. Pachometer results should, however, be calibrated and tested by exposing the rebars for visual inspection in some noncritical locations of the structure.

In addition to on-site tests, laboratory testing is generally required. A determination of compressive strength by testing cores removed from the building is important. The composition of the original concrete can be determined by x-ray diffraction and microscopic analysis. Tests of the reactivity of the aggregate are important to determine if a reaction between cement paste and aggregate has contributed to material deterioration.

Chemical analysis is also important to establish the presence of chloride and in tests for absorption and specific gravity, to determine the porosity and density of the original concrete. This will provide an indication of the material's durability. These tests to determine the composition and characteristics of the original material are essential for the formulation of a compatible design mix for the repair materials.

Moscow subway. Contrast this concrete work with that of the precast building system (P-210). (Photo by F. Wilson)

On the basis of visual inspection, on-site and laboratory testing, and the compilation of background information, an analysis and evaluation of the condition of the structure can be prepared. This must include an assessment of the structural capabilities and conditions of the existing building. It is essential if the impact of deterioration is to be properly assessed and potentially dangerous structural condition identified. This assessment is also essential if the use of the building is to be altered.

TESTING OF CONCRETE

The techniques available for evaluating concrete properties make it possible to be selective in determining which of the components of a concrete structure need repair or replacement and which are sound. The ability to evaluate and rehabilitate concrete building elements selectively has proven a significant factor in encouraging rehabilitation.

The following is an overview of available methods that can be used to evaluate the concrete components of a building. Some of the techniques can be used easily with little or no special training. Others are complex and require a specialist to conduct the test and interpret the results. The tables provide a guide for the selection of the test method available and its parameters.

Nondestructive Tests

Nondestructive testing evaluations:

1. Estimate the properties of concrete material such as composition, variations, hardness, strength, modulus of elasticity and integrity
2. Detect harmful defects in the concrete such as cracks, seams, voids porosity, nonbonds, and inhomogeneties
3. Determine the physical dimensions of the concrete as well as location and size of reinforcement.
4. Can be used to determine conditions of internal stress without damage to the specimen.

This information is obviously useful in determining the condition of concrete members after they have been used for a long period of time. It also is helpful in monitoring concrete to detect signs of failure. In many cases, such as the investigation of cracks in concrete, nondestructive tests may be the only reasonable means of determining the extensiveness of the damage.

A selection of the most appropriate and effective method for nondestructive evaluation of the concrete requires sound judgment based on the information and the cost of the evaluation. Generally, on-site nondestructive or destructive testing should be preceded by a visual inspection of the concrete in all accessible parts of the structure and is used to evaluate further the effects of stresses already observed.

CONCRETE TEST METHODS

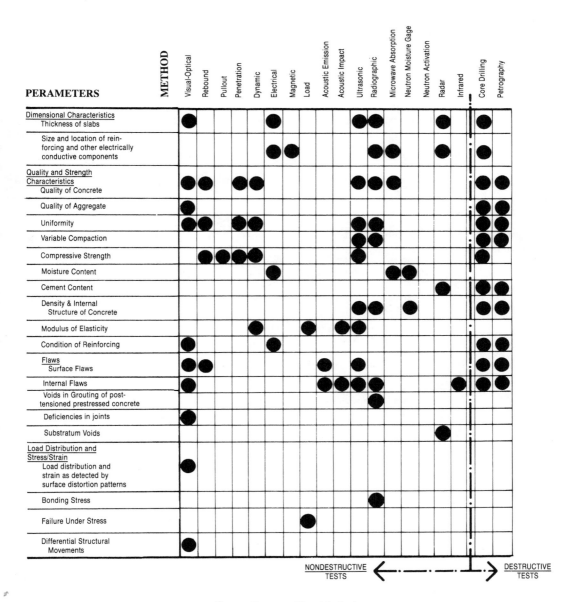

Chart—Concrete Test Methods.

Concrete Test Methods

METHODS	CAPABILITIES	DEGREE OF SENSITIVITY	APPLICATIONS	ADVANTAGES	LIMITATIONS
VISUAL/ OPTICAL					
Visual	Detection of surface laws	Detection of cracks 2–3 microns wide on a smooth surface (smaller with a handheld magnifier)	Concrete samples or in-situ concrete structural components	Inexpensive because no special or power equipment is needed. Can yield defects not detectable by other methods	Only surface information is given. Relationship between surface appearance and condition of total sample must be determined
Surveying, vertical and horizontal movement	Measures differential movements with time	Structural movement may be detected within the precision of the surveying equipment	Long-term observations to determine critical movements of concrete structures	Cyclical relationships between deformation and temperature or load can be derived	Immediate data interpretations cannot be made because of the long-term, cyclical observation periods. A trained surveyor is necessary for data collection and evaluation
Joint survey	Checks for a variety of joint conditions which may indicate problems with concrete, such as spalling; D-cracks; absence, excesses, and condition of joint fillers; and chemical attack	Depends upon the manual measuring device used (calipers, etc.)	Expansion, contraction of construction joints in concrete	An initial step of a more in-depth investigation of concrete problems. Inexpensive because survey is limited to visual inspection and manual measurements	This method is most applicable to foundation walls and slabs. A trained evaluator is necessary for data collection and evaluation
Fiber optics	Detects internal cracks, voids, or flaws if path to surface is available	Working length 30–1330 mm, depending on equipment used	Can be used to look into cracks or areas where cores have been removed	Clear, high-resolution image of remote inspection subjects	Many boreholes are required to give adequate success; expensive

Concrete Test Methods (continued)

METHODS	CAPABILITIES	DEGREE OF SENSITIVITY	APPLICATIONS	ADVANTAGES	LIMITATIONS
REBOUND					
Schmidt rebound hammer	Measures surface hardness. Useful for determining relative quality of concrete. Strength properties of concrete with rebound distance	Calculation of strength from calibration curve. Sensitivity of calculation depends upon accuracy of curves	Concrete samples or in-situ concrete structural members	Inexpensive, fast, and can be operated by laymen	The indication of concrete strength is not accurate, results are affected by condition of concrete surface. Requires a correlation curve between rebound value and concrete
PENETRATION					
Windsor Probe	Measures depth of penetration. Compressive strength is correlated with penetration	Depends upon the location of the test and the precision of the calibrated depth gage	Concrete samples or in-situ concrete structural components	Equipment is simple, durable and can be used by field personnel with little training	Slightly damages a small area of concrete that is 25–50 mm in diameter. Does not provide accurate determination of strength without a correlation between depth of penetration and concrete strength
ELECTRICAL					
Dielectric	Can determine the varying degrees of moisture content of concrete by its insulating capabilities	Has an accuracy of ±0.25% moisture content	Has been used in the past only in laboratory	The equipment is readily automated	Equipment is very expensive and is capable of testing only for moisture content
Electrical resistivity	Can determine slab thickness and rebar location by measuring electrical resistance through the concrete		Applied to in-situ pavements and slabs	Equipment is simple and easy to use	Method is limited to testing of pavements and on-grade slabs and has only specialized applications. Results are inaccurate and are affected by air entrainment, concrete density, moisture and salt content, and temperature gradients

Concrete Test Methods (continued)

METHODS	CAPABILITIES	DEGREE OF SENSITIVITY	APPLICATIONS	ADVANTAGES	LIMITATIONS
MAGNETIC					
Cover meters and pachometers	Used to locate ferromagnetic and electronically conductive components in concrete (both location and depth below surface)	Can detect ferromagnetic components only within 180 mm of concrete surface	Concrete samples or in-situ concrete structural components	Light, portable equipment; easy to operate and relatively inexpensive	Portable battery equipment will not operate satisfactorily in freezing temperatures. Good results are obtained only if one layer of rebars is present. Will not work well with mesh reinforcement
ACOUSTIC					
Acoustic emission	Monitoring of high-frequency acoustic signals (stress waves) leads to detection of growing internal flaws, usually cracking		Concrete samples or in-situ concrete structural components	Equipment is simple and easy to operate. Data gathering requires minimal training	Interpretation of results requires an expert. Background noise distorts results. A computer is recommended for triangulation of flaw location. Very expensive. Can be used only when the structure is loaded and when flaws are growing
Acoustic impact	Measuring of impact energy to detect and evaluate debonds, hairline cracks, and voids		Concrete samples or in-situ concrete structural components	Equipment used is portable, easy to operate, and can be automated	Used mainly for pavements or slabs-on-grade. Most of the equipment is expensive, is in the developmental stage, and is not commonly used

METHODS	CAPABILITIES	DEGREE OF SENSITIVITY	APPLICATIONS	ADVANTAGES	LIMITATIONS
HIGH-ENERGY ULTRASONICS	Evaluating the thickness, quality, and uniformity of a concrete by measuring the velocity of a high energy ultrasonic pulse	Thickness measurements of concrete accurate to ±5.0%	Concrete samples or in-situ concrete structural components	Measurements are very accurate. Currently, it is the only method to measure slab thickness accurately and nondestructively	Large and heavy power supply equipment is required; data interpretations are limited to thickness measurements. Both surfaces of the concrete must be accessible
RADIOGRAPHICS					
X-ray	Density and internal structure of concrete, location of rebars and bonding stress points		Can use only on concrete lab samples	Both radiographic methods provide a permanent record of problems on film	Radiographic equipment is very heavy and expensive for field use with concrete. Both radiation
gamma ray	Location and condition of rebars, voids in concrete and grouting, determination of density and thickness of concrete		Can only be applied to concrete samples or in-situ concrete structural components		sources are injurious to organic tissue and the operators must be adequately shielded. Both surfaces of the concrete must be accessible
MICROWAVE ABSORPTION	Application of the principle of microwave absorption to determine moisture content	Yields values of moisture content within 30% of the mean	Concrete samples or in-situ concrete structural components	Easy to use; moderately priced	Low degree of accuracy; two opposite faces of specimen must be accessible
DYNAMIC OR VIBRATION					
Ultrasonic pulse velocity, and resonant frequency	Measures travel time of ultrasonic pulses through concrete to determine quality and strength (modulus of elasticity, rigidity, and durability) of concrete	The estimations of uniformity and continuity are very qualitative in nature and cannot be discussed in terms of degree of sensitivity	Concrete samples or in-situ concrete structural components	Excellent for determining concrete uniformity	Skill is required to analyze results. Does not provide an estimate of strength. Equipment is expensive and requires field calibration. Background vibrations can affect results

Concrete Test Methods (continued)

METHODS	CAPABILITIES	DEGREE OF SENSITIVITY	APPLICATIONS	ADVANTAGES	LIMITATIONS
NUCLEAR					
Neutron scattering	Measurements of decreased neutron energy results in an evaluation of moisture content	Adequacy of this method for field application on buildings has not yet been proven	Concrete samples or in-situ concrete structural components	An approximate method for measuring moisture content	Equipment is very sophisticated and expensive and is not widely used. Calibration procedures have not been standardized as yet
Neutron activation analysis	The cement content in concrete can be estimated by comparing the neutron activity of the test sample with a reference standard	Accuracy is questionable. Performance of this test method in the field has yet to be proven	Concrete samples of in-situ concrete structural components		Equipment is expensive and complex. Calibration procedures have not been standardized as yet
INFRARED					
Infrared test	Various passive heat patterns are identified with defects such as internal flaws, voids, and growing cracks	At this stage of development, results are relatively unreliable	Concrete samples or in-situ concrete structural components	Has the potential to become a relatively inexpensive and accurate method of detecting concrete defects	Not very reliable; subject is being researched
LOAD					
Load testing	Application of a design load (lead, concrete, or water) to a concrete structural system to verify load carrying ability		Applied to in-situ concrete structural systems	Provides a high degree of reliability on a structure's ability to perform under normal loading	Validity for long-range performance is questionable. May cause cracks, distortion, or even premature failure. Also requires large amounts of preparation and cleanup time
RADAR	Detection of substratum voids	80% reliability of void detection	Concrete samples or in-situ concrete structural components	Far less destructive than "guess and drill" methods; and scanning of large surface areas can be done quickly	Not reliable with slabs containing reinforcing mesh; very expensive; operator needs technical training

METHODS	CAPABILITIES	DEGREE OF SENSITIVITY	APPLICATIONS	ADVANTAGES	LIMITATIONS
PULLOUT TEST	Measures in-situ strength of hardened concrete	Comparable to pull-out of cast-in-place anchors (ASTM C 900)	In-situ concrete structural components	Fast, simple, inexpensive. Easy to apply in the field. Offers direct determination of strength parameters	Within-test variations can be expected to occur because of lack of standardization of test procedures and equipment. Design of split-sleeve assembly is critical. Epoxy grout must cure at least 24 hours before commencing test

REFERENCE

Selected Methods for Condition Assessment of Structural, HVAC Plumbing and Electrical Systems in Existing Buildings; NBSIR 80-2171. Frank H. Lerchen, James H. Pielert, Thomas K. Faison.

A good general procedure to follow in order of priority is to use (1) visual evaluation, (2) nondestructive testing, (3) destructive evaluation, and lastly (4) load tests. Past photos and written records which could reveal changes in the condition of the concrete should be reviewed when available. In-place testing can be used to correlate strength and condition estimates already derived from visual inspections.

Nondestructive testing methods cannot be expected to yield a value for strength, although developments in metal research are beginning to indicate promise of this possibility. The methods that follow attempt to measure other properties from which the evaluator can estimate the strength and related parameters of the concrete. They are summarized in the accompanying chart and table.

Visual/Optical Inspection

Surface Flaws, Finish, Roughness, Scratches, Cracks, Color

Visual inspection is one of the most widely used methods for evaluating surface conditions of concrete. It can be performed with or without the use of optical aids such as low-powered magnifiers. Defects such as missing components, cracks, erosion, corrosion, and misaligned joints, etc., can often be detected with the naked eye. In addition, with the use of five- to ten-power magnifiers, even surface flaws as small as a few microns wide can be detected.

Advantages and Limitations. Obviously inexpensive and requires little or no special equipment. It can also be used to find defects which otherwise cannot be found. However, visual inspection can be time consuming and is absolutely dependent upon the visual acuity and experience and training of the inspector. No subsurface information can be obtained using this technique. A correlation must have been established between surface conditions and serviceability of the concrete.

ASTM Standard Practice for Examination and Sampling of Hardened Concrete in Construction (C-823-75, reapproved 1981)

In visual inspection of concrete ASTM Standard C-823 suggests the following: The examination should locate and describe all of the categories of concrete. All affected constructions or portions should be identified and the external aspects of failure described and quantitatively as possible. Photographs of pertinent features of the construction, their environs, and the manifestations of failure are valuable and should be obtained.

Observations. Features to be noted:

1. The nature and extent of cracking and fractures
2. Evidences of volume change, deflection, or dislocation of the constructions or portions of it; which may include the closing or opening of joints, tilting, shearing, or misalignment of structural elements and shifting or misalignment of machinery
3. The condition of exposed surfaces, especially such features as spalling, popouts, unusual weakness, disintegration, excessive wear, and discoloration
4. Evidences of cement-aggregate reactions
5. Secondary deposits on surfaces, in cracks, and in voids
6. The presence and extent of repair work and the quality of its bond to the original concrete.

Some of the other properties of the concrete requiring observation include: the thoroughness of consolidation; whether the concrete is air-entrained; evidences of segregation and bleeding; indications of extremely high, low, or normal water content; in the case of reinforced concrete, the condition of the steel and its location in the section; and the nature and condition of other imbedded items.

Any phenomena indicating distress of the concrete should be studied in relation to possible causative or contributory factors, such as varying conditions of exposure over the area of the constructions; the sequence of placing operations; conditions prevailing during construction; sources of supply of concrete and concrete-making materials; identifiable problems of handling, placing, and finishing; condition of curing and early protection; and the adequacy of the structural design and conformance to the plans.

Varying conditions of exposure over the area of construction during and after construction may include the following:

1. Differences in thermal exposure to solar heating. Shaded portions probably are subjected to the lowest range of diurnal thermal cycles.
2. Differences in exposure to moisture. These may arise by orientation of the construction with respect to prevailing winds during times of rainfall or snowfall, and will be affected by the diurnal thermal cycles.
3. Differences in the mineral composition of the subgrade, so that part of the construction is located on a foundation containing swelling clay or containing unstable sulfides or sulfates.
4. Differences of the moisture content of the subgrade during or after construction.
5. The foundation and subgrade materials and conditions should also be carefully examined if there is a possibility of their involvement in serviceability of the concrete.

Surveying—Horizontal and Vertical Movement

Different Structural Movement

Differential measurements of structural movement can be made by making use of horizontal and vertical control standards. Survey points are established at various key areas of the structure, and their movement in both horizontal and vertical directions is measured over extended time intervals. These movement measurements can be compiled into history plots that may cover months or even years. The history plots may reveal normal cyclical movements associated with temperature, moisture condition, or applied load, and may also reveal some type of structural deficiency. The observed movements must be interpreted relative only to the structure under observation, and only those movements that exceed the expected values should be considered as a source of deficiency.

Advantages and Limitations. This method requires a long period of time to compile enough data to make accurate judgments of the structure's stability. Many times an evaluation of the structure is needed quickly, which makes the surveying method impractical. The method requires trained personnel to operate the surveying equipment and to gather data. Surveying is very useful only when some form of structural distress is suspected and structural movement is believed to be the cause.

Joint Survey

Deficiencies in Joints

A periodic inspection of all expansion, contraction, and construction joints should be made when considering the condition of concrete structural members. It must be determined whether or not each joint is in good condition and functioning as designed. Consultation of construction drawings may be necessary for a complete listing of the joints and their designed function. The inspection of each joint should include an examination for spalling or D-cracking (cracking along a joint in a D-shape), evidence of seepage, chemical attack or erosion. Openings in each joint should be measured. A measurement of surface offsets on either side of the joint should be included in the inspection procedure.

Advantages and Limitations. The joint is an inexpensive and fast method of assessing the quality of the various types of joints. Since joints are essential in the construction of all concrete structures, there exists a probability that at least some of these joints are not functioning as designed and may be a source of some type of concrete deficiency. A trained inspector is necessary to complete a thorough inspection. Caution should be taken to evaluate the entire joint when making the final decision as to the overall condition.

Fiber Optics

Detection of Internal Cracks, Voids, or Flaws

The innovation of fiber-optic inspection techniques came from the need for remote visual inspection of areas accessible only through small-diameter openings. The principle of fiber optics makes use of a 1.0–2.7 mm (0.04–0.11 in.) diameter single glass fiber with a graded index of refraction surrounded by flexible, light-conducting fiber bundles that are approximately 20 microns in diameter. The larger glass fiber acts as a lens to transmit the remote image that is illuminated by the light-conducting fiber bundles. The entire fiber assembly is covered by a stainless steel sheath to protect it from damage in use. The probes range in working lengths from 30 to 1300 mm (1.2 to 52 in.). The probe is attached to a portable halide or cold light source that is adjustable and provides 150 watts of concentrated white light to the fiber bundles. The application of this method to a concrete structural member involves inserting the probe into a crack or drilled hole and looking through the eyepiece for various flaws such as cracks, voids, or aggregate debonds. The fiber-optic apparatus is adaptable to photographic and television equipment for permanent records of inspection. It is most commonly used to look into cracks in areas where cores have been removed.

Advantages and Limitations. Fiber optical equipment can provide clear, high-resolution images of remote inspection subjects. This equipment is also easy to handle and operate. However, its applications to a material such as concrete are limited. A complete visual inspection of a concrete member using fiber optics requires a large number of boreholes, which might weaken the member. The equipment is also quite expensive (about $150 per probe), and many times more than one size of probe is necessary to complete a detailed inspection. Fiber-optic equipment is best applied to inspections of machined parts or other mechanical assemblies, but can also be used, when necessary, for inspection of small areas of a noticeably deficient structural member.

Rebound Test

Quality and Uniformity, and Compressive Strength Estimates

An instrument which can be used to estimate the compressive strength of concrete by measuring the height of rebound of a hardened steel hammer dropped on the concrete is the test hammer developed in 1948 by Ernst Schmidt, a Swiss engineer. The Schmidt rebound hammer consists of a spring-activated hammer and plunger mounted within a tubular housing. To operate the hammer, the impact plunger is placed against the surface of the concrete specimen and the hammer housing is pressed down until the plunger almost disappears into the housing. The hammer then will release itself. A built-in scale measures the rebound value in percent of the forward movement of the hammer mass after it strikes the concrete surface. The amount of rebound is affected by many factors, such as the composition of the concrete, aggregate properties, surface texture and hardness, moisture content, and the mass of the concrete specimen. User's charts (provided with the equipment) are based on empirical correlations which have been established between strength properties, hardness, and the rebound number. ASTM Standard C805 "Test for Rebound Number of Hardened Concrete" describes this test method.

Advantages and Limitations. The Schmidt hammer provides the advantages of being able to compare quality of concrete from different areas of the specimen, thereby detecting areas of potentially low strength. Since the equipment is so lightweight, simple to operate, and inexpensive, it is suitable for both laboratory and field use.

Use of published calibration data to estimate strength of concrete from rebound readings is of only limited value and is not recommended because of the number and nature of the factors affecting the rebound. Therefore, the user should use strength estimates derived from this test method with extreme care. It is advisable to calibrate the test equipment frequently with the strength of cores drilled from the structure. By doing so, the user will be able to determine the degree of reliance that can be placed on the strength estimates from the rebound readings.

The user of this test should be aware that no correlation has been determined to exist between rebound readings and modulus of elasticity. Its use should be limited to determining the quality control and uniformity of concrete and estimating the compressive strength with a maximum accuracy of $\pm 20\%$. Even then, these accuracies can be expected only if the specimen has been cast, cured, and tested under conditions which were identical to those from which the calibration curves were established.

Windsor Probe Device

Quality and Uniformity, and Compressive Strength

The Windsor probe device consists of a driving gun which uses a powder charge to drive a high-strength steel probe into the concrete to be tested. Generally, three probes are driven in a triangular pattern; and the amount of penetration is determined by measuring the length of the probe extending from the surface of the concrete. Then the depth of penetration measurements can be converted to concrete strength determinations by using calibration curves applied by the manufacturer.

Advantages and Limitations. Windsor probe equipment is simple, durable, requires little maintenance, and can be used by laymen in the field with little training. It is fast—three tests are made in approximately five minutes. ASTM Standard "Tests for Penetration Resistance of Hardened Concrete" (C 803) describes this test method.

The Windsor probe primarily measures subsurface hardness and does not yield precise measurements of the in-situ strength of concrete. The probe test is, however, useful in assessing the quality and relative strength of concrete. As with the Schmidt hammer (rebound) test, interpretation of test results depends on other known factors which are based on correlation plots. When used in concrete which is 40—50 years old, this test method may yield a higher strength than actually exists.

The Windsor probe test does minor damage to the concrete over an area approximately 25–50 mm (1–2 in.) in diameter, leaving holes about 8 mm (0.3 in.) in diameter for the depth of the probe. It may also cause minor cracking, necessitating superficial repairs.

Care should be taken to insure that the user observes normal safety precautions concerning its use and handling.

Electrical Tests

Moisture Content

The technique of using dielectric measurements to determine the moisture content of concrete is based on the principle that the dielectric properties of concrete change with change in its moisture content. In this test, the dielectric testing equipment is placed on the specimen, the oscillator is energized, and the overall capacitance of the specimen is measured between two coplanar electrode plates. The two electrode plates are separated by insulation to prevent short circuiting and to ensure that the measurement is of the current traveling through the specimen.

It has been determined by developers of dielectric measurement equipment that the best frequency range to use when measuring the dielectric constant is 10–100 MHz. Buy using these high frequencies, the effects of conductance caused by dissolved salts and faulty contacts with the electrodes are greatly minimized.

Advantages and Limitations. The greatest advantage to using dielectric measuring equipment is that it is readily automated. However, dielectric measuring equipment is very costly because it uses high-frequency currents. This limits its use to only those specialized applications which can justify the expense. The test method is further limited by the fact that it can test only for moisture content of the concrete. Successful use of dielectric techniques is contingent upon careful and frequent calibration of the measuring instruments.

Another limitation to consider carefully is the fact that the dielectric properties of concrete are highly dependent on salt content and temperature of the specimen as well as moisture. This can pose problems in the interpretation of test results.

Electrical Resitivity Test

Thickness of Slabs and Location of Reinforcing Steel

Electrical resistivity tests are based on the principle that different materials offer different degrees of resistance to the passage of electricity. Thus, a concrete pavement has a resistivity characteristic which differs from that of the underlying subgrade layers; and this change is resistivity is measurable with electrical equipment. In this test, four electrodes are placed in a line on the concrete specimen at equal distances. A voltage is applied across the two electrodes. For practical purposes, the penetration of the applied current is considered to be equal to the spacing between the electrodes.

After testing, resistivity is computed by using the following formula:
$$P = 2SE/I$$
where S = Spacing between electrodes, in cm
E = Current flowing between outer two electrodes, in amperes.

Resistivity is then plotted against electrode spacing (or depth) and slab thickness, and the location of reinforcing steel is determined from the curves.

Advantages and Limitations. Although electrical resistivity test methods and equipment are simple and easy to use, their one major drawback is that they are limited to testing of concrete pavements. Furthermore, the test results are affected by air entrainment, by mix proportions and density of the concrete, and by the presence of steel reinforcement, moisture, salt, and/or temperature gradients. Hence, the effectiveness of the test results may be questionable. Another limitation to the widespread use of the test method is the high cost of the generators required to produce the high-frequency currents needed to satisfactorily determine the desired results.

Magnetic-Electrical Field Evaluation

Location and Size of Internal Reinforcement or Other Electrically Conductive Components

The magnetic-electric test is based on the principle that ferromagnetic and electrically conductive materials affect the field of an electromagnet. It is used primarily to detect ferromagnetic components, such as rebars in concrete, but also is effective in locating electrically conductive components in masonry and other nonconductive materials.

Several portable, battery-operated, magnetic electric field test devices are now commercially available. Generally, those that measure the depth of reinforcement cover in concrete are known as cover meters, and devices that measure both the cover and the size of reinforcement bars are called pachometers. Typically, these devices are light and hand operated.

The devices are easy to use and require only that the user place the attached probe device over the reinforced concrete and move it about until the reinforcement is determined to be parallel to the length of the probe, at which time a reading is obtained on the dial of the instrument. For a rebar of known diameter, the dial gives a direct reading of the cover to steel.

Advantages and Limitations. The biggest advantage to using the magnetic-electric field evaluation method for detecting internal reinforcement and electrically components is its portability of operation. It is limited, however, to detecting concealed ferromagnetic and electrically conductive components located within 180 mm (7 in.) of the surface of a nonconductive base specimen. Also, in heavily reinforced sections, the effects of secondary reinforcement cannot be eliminated, and the cover of the steel cannot be determined accurately.

An operational characteristic that could prove to be a drawback (depending on geographic location) is the fact that the battery pack for portable units will not perform satisfactorily in temperatures less than 0°C (32°F).

Acoustic Emission Evaluation

Structural Integrity (Cracks, Non-Bonds, Inclusions, Interfaces, etc.)

Acoustic emission evaluation consists of monitoring and evaluating high-frequency acoustic signals (stress waves or pulse) which are produced naturally by the test materials themselves when placed under stress. These acoustic signals are related to the internal physical changes taking place within the specimen being tested. Detection of the stress waves is accomplished by affixing sensors to the specimen, amplifying the acoustic emissions, and recording them on a tape recorder or processing them through a digital computer for recording and analysis.

The technique is based on the Kaiser effect—the principle that, under a load almost all materials produce a significant level of small-amplitude elastic stress waves above the base level of stress waves produced without a test load (unless a flaw occurs or the original load is exceeded). A high emission rate above the base level of emission at operating load, design load or test load level indicates that a flaw is growing.

Advantages and Limitations. Although the equipment is relatively portable and easy to operate, the interpretation of the results requires an expert who has considerable experience in this specialized field. Detection of crack growth is only the beginning of the evaluation. Determining the location of a crack with this equipment is very complex and often requires the use of a microcomputer (for high-speed triangulation). This test is very advantageously combined with load testing.

A significant disadvantage to this method of crack evaluation is the large expense involved. It is the type of test that is most appropriately accomplished by contracting the service from a testing company that specializes in nondestructive testing. Also, the reliability of the detection of flaws is affected by noise sources other than structural defects. Research is currently going on to develop methods for noise source identification and discrimination to minimize this problem.

Acoustic Impact Evaluation

Integrity (De-Bonds, Cracks, Voice, etc.) and Modulus of Elasticity

Acoustic impact evaluation equipment utilizes the principle that the acoustic and impact energy transmission characteristics of a material are related to the geometry and mass effects (e.g., shape, density, and structure) of the material. As compared to the characteristic energy transmission patterns for an unflawed specimen, these patterns will be changed by flaws or changes in the internal structure of the specimen to be evaluated. In other words, a specimen with flaws, (de-bonds, hairline cracks, voids, etc.) will dissipate acoustic energy more readily than a specimen in sound condition.

For certain cases, these energy-transmission patterns can be detected audibly, amplified electronically, and displayed on an oscilloscope, a computer-interfaced cathode ray tube (CRT) or graphic printer, or a voice print spectograh.

Advantages and Limitations. Acoustic impact evaluation equipment is mounted on a trailer, is easy to operate, and can be automated. It provides a method of detecting and evaluating de-bond, hairline cracks, and voids in specimens (although its reliability has not been demonstrated clearly to date). Either acoustic or rebound energy sources can be used to stimulate the test energy pulse which is measured.

It should be realized that before test results can be interpreted, a standard acoustic pattern must be determined for each given application or specific material. Also, when using acoustic impact sources, acoustic isolation is required because of the possibility of outside acoustical interference with test results.

A disadvantage of this test method is the expense of the equipment. It is still in the developmental stage and is used mainly by the Department of Transportation for determining properties of pavements. A potential use is seen in slabs-on-grade for structures.

High-Energy Ultrasonics

Thickness, Quality, and Uniformity

Measuring thickness and estimating the quality of concrete by the high-energy ultrasonic evaluation method is accomplished by measuring the travel time of a large unit of acoustic energy driven through the concrete specimen. A high-energy ultrasonic source (one million electron volts energy with pulse duration of a few nanoseconds) is required to produce measurable ultrasonic indications, since the signal from a standard ultrasonic unit is scattered through the concrete so completely that no information from the reflected signal is available.

Advantages and Limitations. Some manufacturers of this equipment claim that high-energy ultrasonics can measure thickness of concrete masses with an accuracy of ±5%. This technique is limited to measuring thickness of only heterogeneous materials such as concrete. Also, the information received as a result of the test is limited to thickness measurement and does not include information on internal flaws.

Another limitation to the use of this equipment is the weight and size of the electrical power supply. Although the power supply is considered to be portable, available products weigh approximately 91 to 181 kg (200 to 400 lbs.).

Radiographic Evaluation

A. Using X-Ray

Density and internal structure of concrete, size and location of reinforcement, and bonding stress

B. Using Gamma Ray

Location, size and condition of rebars, voids in concrete, voids in grouting or post-tensioned prestressed concrete, variable compaction in concrete up to about 6 m (20 ft.) thick, density of structural concrete members, thickness of concrete slabs, and density variations in drilled cores from concrete road slabs.

Radiograph evaluation techniques are based on the principle that the rate of absorption of x-rays or gamma rays is affected by the density of the test specimen. X-rays or gamma rays are emitted from a source (x-ray tube, radioactive isotope, etc.), penetrate the specimen, and exit the opposite side of the specimen, where there they are recorded on a sensitive film as radiographs (pictures) or viewed on an electro-optical imager (screen).

Flaws having a diameter as small as about 1% of the material thickness (in the direction of the radiation penetration) can be detected in a wide range of materials. Orientation of the defect (parallel or perpendicular to the beam) will affect detection of the flaw. Voids or relatively low-density areas of the specimen allow a relatively higher rate of passage of the rays, which will appear as dark indications on the film (as compared to the lighter background). The reverse holds true for areas of higher density, such as steel reinforcement, which are clearly detected and recorded on the film.

Advantages and Limitations. Radiographic evaluation techniques offer the advantage of providing a permanent record on film for inspection, or a "live" image on an electro-optical system.

Since x-ray evaluation equipment for concrete evaluation is heavy and costly it has very limited application as a field test. It lends itself better to application in a laboratory environment. On the other hand, gamma ray equipment is easily portable because sources (such as cobalt 60, iridium 192, and cesium 137) do not require electrical power. Thus, gamma rays are becoming increasingly accepted as a viable field test method. Gamma radiography can be carried out only by personnel who are licensed by the Nuclear Regulatory Commission. A word of warning: x-ray and gamma ray sources are harmful to organic tissue. The test environment must be shielded (or isolated) as necessary to prevent radiation hazards. Lead shielding required for gamma ray sources can be very heavy, sometimes weighing up to 227 kg (500 lb.).

Other drawbacks to the use of radiographic evaluation equipment are the high initial cost of the equipment and film, the time delay required to develop the film, and the requirement of access to both sides of the test specimen to set up and use the equipment.

Microwave Absorption Test

Moisture Content and Quality, Rebar Size and Location

Microwaves are electromagnetic wave emissions which lie between television and infrared emissions in wavelength and frequency. Because they are electromagnetic in nature, they can be reflected, diffracted, and absorbed by different materials at differing rates. Water, for example, absorbs microwaves at a much higher rate than concrete. Microwave testing methods employ this principle of absorption to measure the moisture content and quality of concrete and other porous building materials and to measure size and location of reinforcement.

The testing apparatus consists of a portable 3,000-MHz (100-mm-wavelength) radio transmitter which is modulated at 3 kHz by a square wave and a 3 kHz receiver which has attached to it an attenuator calibrated in decibels. The transmitter and receiver are attached to opposite sides of the specimen and switched on. The receiver attenuator is adjusted to a null reading both with and without the specimen present. The difference between the two null readings then is translated into moisture content by use of a calibrated chart which is developed by the user. Instructions for developing the calibrated chart and for making the test measurements are supplied by the test equipment manufacturer.

Advantages and Limitations. Although microwave absorption testing equipment is easy to use and is moderately priced, the major limitation to its widespread use is its low degree of accuracy and the difficulty of interpreting test results. For meauring moisture content in concrete, the expected accuracy is no better than 30% of the mean value. this low accuracy is due in part to the fact that concrete is a heterogeneous material which causes internal scattering and diffraction of the microwave emissions (scattering is also caused by reinforcement).

Dynamic Tests—Ultrasonic Pulse Velocity and Resonant Frequency

Quality and Uniformity, Strength, Modulus of Elasticity

The ultrasonic pulse velocity test method works on the principle that vibrational wave propagation is affected by the general quality of a dense medium such as concrete. Assuming concrete to be an inherently heterogeneous medium, the path of travel of the ultrasonic pulses must be long enough to test an "average" section of the structural member. The equipment makes use of an electro-acoustic transmitter and a receiver that are placed in contact with the concrete surface. The distance between them is measured to a high degree of accuracy by the travel time of an ultrasonic pulse. The average pulse velocities have been correlated with general concrete conditions and strength qualities in the following table:

Pulse Velocity (m/sec)	Condition/Strength Evaluation
Above 4575	Excellent
3660–4574	Good
3050–3659	Questionable
2135–3046	Poor
Below 2135	Very Poor

The main point to remember is that the less impedance (velocity slowing) that the ultrasonic pulse experiences, the higher the quality of the concrete.

The resonant frequency test method is applied almost exclusively in the laboratory and is used to determine the natural frequency of vibration of a standardized concrete prism or cylinder. The vibration frequencies found in the test area are used in turn to calculate Young's modulus of elasticity and rigidity, Poisson's ratio, and durability ratings. A simplified description of the testing process is as follows: (1) a drive unit applies standardized mechanical vibrations to a concrete sample; (2) the vibrations are picked up by piezoelectric sensors; and (3) the vibrations are fed into a cathode-ray oscilloscope that give a digital and graphic readout of the natural resonant frequency of the sample.

Advantages and Limitations. Both test methods yield an accurate assessment of concrete quality and uniformity. Strength then can be estimated by preparing calibrated charts. The accuracy of estimation is ±15–20%. Both methods are applicable only to laboratory tests and require a trained operator to evaluate the results. They are in frequent use, and the equipment is relatively inexpensive to purchase.

Neutron Scattering Methods (Neutron Moisture Gages)

Moisture Content

The neutron scattering method for determining moisture content in concrete is based on the principle that hydrogen-containing materials (such as water) decrease the speed of fast neutrons (greater than 0.1 Mev) in accordance with the amount of hydrogen produced in the test material. In the case of concrete, the moisture in the concrete is the major source of hydrogen and calcium hydroxide.

Thus the evaluator can obtain an estimate of the hydrogen content of the specimen by counting the slowed-down neutrons which result from interaction with the hydrogen. Most neutron moisture gauges use isotopic neutron sources.

Advantages and Limitations. Although neutron moisture gauges are satisfactory for measuring moisture contentr in concrete, they have not been used widely since procedures for calibration and measurements have not been adequately standardized. One of the biggest drawbacks is the fact that gradients of moisture content near the surface of the test specimen and the presence of neutron absorbers within the specimen can cause erroneous results. The adequacy of this test method for field application has yet to be proven.

Neutron Activation Analysis

Cement Content

The neutron activity method of nondestructive evaluation is based on the principle that concrete becomes radioactive when bombarded by neutrons. The unstable radioactive isotopes that are formed emit beta and/or gamma radiation which allows a clear characterization of the radiosotope and its arent nuclide.

To perform the test, a reference standard (of the element being determined) is activated under the same conditions as the unknown sample. Then the weight of the element in the sample under investigation can be determined by comparing the activities induced in both the test sample and the reference sample.

The major use of neutron activation analysis techniques for evaluating concrete is to determine the cement content of concrete. Since the neutron activation process is independent of chemical bonding and the internal physical condition of the specimen, these respective properties cannot be measured using this method of evaluation.

This test is not widely used for several reasons, the first of which is that it is expensive and complex. Since it is a recently developed test method the procedures for calibration and measurement have not been standardized adequately. Then too, the accuracy of the test results is questionable because gradients of moisture content near the surface of the concrete and the presence of neutron absorbers within the concrete will induce errors in the test measurements. Performance of this test method in the field has yet to be proven.

Infrared Test

Internal Flaws, Crack Growth, Internal Voids in Slab

Infrared evaluation techniques detect flaws in concrete by using selective infrared frequencies to detect various passive heat patterns which can be identified as belonging to certain defects. There are three commonly used variations of this technique of nondestructive evaluation of materials—one based on steady-state heat patterns, one based on transient heat emission, and one based on low-amplitude vibrations.

1. *Steady-State Heat Pattern Method.* In this method of infrared evaluation, the heat pattern in the specimen is produced by a boron-epoxy laminate during cyclic loading at 30 Hz. The heat pattern is caused by the dissipative nature of

the internal friction inherent in the specimen material. As fatigue damage develops, changes in the heat pattern can be observed.

2. *Transient Heat Emission.* The heat pattern resulting from the transient heat emission method is produced by the dissipative part of the strain energy released when a material fractures. This strain energy released at the time of fracture is partly in the form of heat, which produces the image seen by the infrared camera. For this test, a boron-aluminum laminate is used on the test specimen to produce the heat patterns.

3. *Vibrothermography.* Rather than including a cyclic (fatigue) loading in the specimen, low-amplitude vibrations can be ued to produce heat patterns without the need for a fatigue machine. Standard shaker tables or transducers are used to produce the vibrations needed to conduct the test.

Advantage and Limitations. The greatest advantage of using the transient heat emission method is the ability to monitor quantitatively energetics of fracture events by reviewing images produced on infrared film. The steady-state heat pattern method for detecting flaws in concrete currently is not reliable enough to make its use widespread. It shows great potential and is undergoing research and development. Concrete is a good material to use with these two test methods because it is a good insulator, it has a low specific heat value, and it has high heat capacity.

Vibrothermography has four advantages over the other two infrared test methods: (1) the low level of input energy does not damage the test specimen in any way (2) the equipment is less expensive and easier to use (3) the frequency of the excitation can be varied over a wide range; and (4) often flaws can be selectively discriminated by varying the frequency of the excitation.

Infrared evaluation techniques can be used advantageously in conjunction with load tests.

Load Testing

Performance Under Load (Rigidity, Failure Under Stress)

Load testing verifies design load-carrying ability of a structure (unless failure occurs during the test). The test procedure is based on the principle that any structure which is capable of withstanding the stresses of a design load should perform satisfactorily under actual loading conditions.

To conduct the test, a test load is applied to the structure in a manner that will simulate the load pattern under design conditions. The test failure indications (such as leakage, distortion, or even structural failure) then are monitored and measured visually or with detection devices. Structural loading can be accomplished by the use of heavy dead weights as concrete, lead, or water.

Advantages and Limitations. The advantage of using load testing for nondestructive evaluation of concrete is that it provides a high degree of reliability on the structure's ability to perform satisfactorily under normal loading conditions during the present time. However, its validity concerning the long-range performance of the structure is questionable.

A serious limitation to the use of this test method on old buildings is the fact that it may cause cracks or distortion, or may even cause premature failure of the structure or some of its elements. Load testing has the disadvantage of requiring a significant amount of prepartion and clean-up time before and after the test. Because of its high cost, it has limited application.

Radar Test

Detection of Substratum Voids

Since slab-on-grade is one of the most common construction techniques used in buildings, it is useful to know that an alternative test method is being developed to check the integrity of the subgrade material. Some manufacturers claim that radar, which employs the use of transmitted electromagnetic impulse signals, has an 80% reliability of substratum void detection. This new test method is more time and cost efficient and less destructive than the previously used "guess and drill" techniques.

Advantages and Limitations. Radar equipment is expensive. A major disadvantage of the radar system is that the reliability of void detection is greatly reduced if the slab contains reinforcing bars or mesh. In such cases, the data must be interpreted by an expert trained in the field of electromagnetic impulse analysis. This test method may also have further applications in geotechnical and highway engineering, where void detection is important in determining the strength and service life of a concrete slab.

Pullout Test

In-Situ Pullout Strength of Hardened Concrete

A modification of the ASTM C 900 pullout test of cast-in-place plugs has been developed for determining pullout strength of hardened concrete using drilled and set-in-place wedge anchors. The test is based on the determination of the pull force required to cause shear failure of the concrete. The pull force is derived from a tension ram connected to the anchor at one end and a portable hydraulic pump at the other end.

Two types of wedge anchors currently are considered acceptable for use in this test: (1) the split-sleeve assembly, and (2) epoxy grouted bolts. The split-sleeve assembly consists of a tapered steel rod threaded at the small end and a specially designed high-strength sleeve that includes a split shell at one end. The other end of the sleeve is threaded so that it can be screwed to the base plate of a tension ram. To prepare for the test, holes measuring 19 mm (¾ in.) in diameter and 38 mm (1½ in.) deep are drilled in the hardened concrete with special drilling equipment designed to permit rapid drilling of holes perpendicular to the surface of the concrete sample. A description of the remainder of the two-test methods follows.

Using Split-Sleeve Assembly

1. Insert and secure the steel rod and split-sleeve assembly in the drilled hole with tapered end down.
2. Screw the assembly to the base plate of the ram.
3. Gradually apply a pullout force on the rod through the tension ram until the plug breaks out of the concrete.
4. Record the maximum pull reading observed.

Using Epoxy-Grouted Bolts

1. Clean and dry the drilled hole.
2. Fill the hole with a flowing epoxy.
3. Place the anchor bolt (a threaded steel rod) into the epoxy with a slow rotary motion (to avoid the formation of entrapped air at the interface).
4. Hold the bolt in a perpendicular position until the initial set of the epoxy takes place.
5. After proper curing of the epoxy (usually 24 hours), the bolt and resulting concrete plug is pulled out using a tension ram assembly. The load is applied uniformly until failure of the concrete occurs.

The test results of the pullout test on hardened concrete indicates good correlation with the strength results obtained from compressive tests on cylinders and drilled cores. In fact, the correlation coefficient values of around 0.81–0.85 are comparable to those reported from the pullout test of cast-in-place anchors (ASTM C 900).

Advantages and Limitations. The pullout test on hardened concrete is a simple, quick, and inexpensive test which can be carried out easily in the field on in-situ concrete. It offers the advantage of direct determination of some strength parameters and shows a good degree of correlation with the standard strength test (compression test on cylinders or drilled cores). However, due to the lack of standardization of the test procedures and the equipment, high within-test variations may be expected to occur. For this reason, at least six pullout determinations per test should be performed.

For the split-sleeve assembly method, the effectiveness of the test is dependent upon the design and geometry of the split-sleeve assembly itself. Proper design of the split-sleeve assembly is important because it is the mechanism used to transmit the while pullout force to the concrete.

For the epoxy-grouted bolt method, the epoxy must have suitable pot-life characteristics and provide satisfactory bonding action with the concrete. It also must be able to develop high elastic and strength properties at a very early age under variable ambient temperatures. Preliminary use of this test method indicates that research is needed to develop epoxies that will allow the pullout tests to be performed in a matter of hours rather than after 24 hours of set time.

Destructive Tests

Destructive testing, in contrast to the varied nondestructive test methods, are fewer in number. The most commonly used methods for destructive evaluations are listed here.

Destructive test procedures include the test methods of core drilling and petrography, which can be used together to determine the strength and internal quality of concrete and its cement and aggregate components. Such characteristics as deleterious chemical reactivity between components, and physical condition and composition of the cement and aggregate can be determined best by these destructive test methods.

Core Drilling

Strength and Internal Quality

The existence of some surface aberrations such as scaling, leaching, pattern cracking, and freeze-thaw weathering may suggest the need for an investigation of the internal condition of a concrete structural member. A widely accepted destructive test method to check for internal quality is core drilling.

Drilled cores of 50–250 mm (2–10 in.) in diameter and of varied lengths are obtained using a gasoline- or electric-powered drill rig. Cores from older members with rehabilitation potential tend to be 50–150 mm (206 in.) in diameter. The drill bits are either calyx (steel shot) or diamond (bort). The diamond bit drills are much faster and cleaner but require a more skilled operator and are much more expensive. Cores should be cataloged when drilled. They should then be labeled carefully and packaged for transport to a test laboratory. The sample types can be placed in two broad categories: (1) samples that are intended to be representative of the concrete in place, and (2) samples that display specific features of interest, such as a core taken at the site of a cracking pattern.

Subsequent tests performed on the sample will vary significantly. Core size requirements vary with the type of test to be performed. The basic testing list includes: (1) visual inspection (surface condition, depth of deterioration, fractures, unusual deposits, coloring or staining, distribution and size of voids, location of construction joints, and contact with the foundation or other surfaces); (20 compressive and tensile strength; (3) transverse, longitudinal, and torsional frequencies; (4) static modulus of elasticity; and (5) Poisson's ratio of concrete in compression. Other tests that might be warranted are dynamic loading and seismic loading tests. These last two test methods require cyclic loading techniques used with various time intervals. ASTM C 41 "Standard Method for Obtaining and Testing Drilled Cores and Sawed Beams of Concrete" gives a more complete description of this test.

Advantages and Limitations. This is the most widely accepted method to determine strength and quality of in-place concrete. The process of drilling and analyzing concrete cores is expensive and requires a relatively long time period (1 to 3 weeks) for a thorough analysis. This test method is termed destructive because repeated core drilling in a concrete structural member could affect its integrity. Cores obtained from appropriately located drilling sites are a very accurate representation of the internal composition of the concrete. Core drilling is best suited to concrete structural systems such as walls and foundations. It should be used very judiciously when dealing with structures being evaluated for rehabilitation purposes.

ASTM C-283

Specimens of hardened concrete may be obtained by coring, sawing, or otherwise removing portions of the concrete.

Sawing or coring by rotary drilling is preferred for samples to be subjected to tests of physical properties or to petrographic examination. Caution should be used to avoid or to minimize fracturing of the concrete and contamination of the sample with foreign substances. Use of sledges, chisels, and similar tools should be avoided; their possible effects on the integrity of the sample must be considered during examination and testing of samples.

Diamond-drilled cores are preferred to shot-drilled cores because the outer surfaces of diamond-drilled cores are smoother and reveal the composition and fabric of the concrete more clearly.

If feasible, samples should be taken perpendicular to the layers in which the concrete was deposited. The sample should include the exposed surface, near-surface concrete, any concrete in contact with aggressive waters or other aggressive substances, and concrete at depth.

Samples of isolated spalls or popouts should include representative examples of the spalls and popouts and the underlying and adjacent concrete.

Petrography

Physical Quality of Aggregate and Concrete

Petrographic examination uses microscopy and/or x-ray diffraction and differential thermal analysis in conjunction with core testing to aid in determining the physical and chemical properties of aggregates and concrete.

1. Aggregates

By the use of this test, the aggregate is identified according to mineralogical and chemical differences, and the following properties of the aggregate may be determined:
  (a) The extent to which the particles are coated
  (b) The nature of the coating substance
  (c) Particle shape
  (d) Potential deleterious reactivity with cement alkalis
  (e) Chemical reactivity
  (f) Condition of the aggregate
  (g) Composition of fine and coarse aggregate.

To accomplish this test of aggregate found within a concrete core sample, the aggregate is examined by hand-held lens or a microscope. The aggregate then is analyzed in accordance with the following three criteria.

Petrographic Identity. An experienced petrographer can examine the aggregate samples to identify whether the particles are coated with either mineral or organic coatings, to identify the coatings, and to determine the potential physical and chemical effects of the coating on quality and durability of the concrete. The particles are examined further to determine their classification (type).

Physical Condition. Aggregate samples are examined to determine their shape, flatness, angularity, and other pertinent properties that can lead to classification of the aggregate as: (1) satisfactory (contributes to high or moderate strength, abrasive resistance, and durability of concrete); (2) fair (contributes to moderate strength, durability and abrasive resistance under ideal conditions, but concrete might break down under rigorous conditions); and (3) poor (contributes to low strength and poor durability of concrete or breakdown of concrete).

Anticipated Chemical Stability in Concrete. The amount of deleterious aggregate particles (alkali reactive rock types such as opal) may be determined. The deleterious particles can be responsible for producing adverse effects on the concrete through chemical reactions with the cement alkalies.

2. Concrete

The condition of the concrete also may be evaluated by using petrography. For this purpose, the most useful samples of concrete are diamond-drilled cores with a diameter of at least two to three times the maximum size of the coarse aggregate, used in the concrete. For concrete with 150-mm (6-in.) aggregate, a core of 200–25) mm (8–10 in.) diameter is commonly used rather than the more costly and harder-to-handle 305–457 mm (12–18 in.) cores prescribed by this general rule-of-thumb.

The following features of concrete can be studied and evaluated:

(a)   Denseness of cement paste
(b)   Homogeneity of the concrete
(c)   Occurence of settlement and bleeding of fresh concrete
(d)   Depth and extent of carbonation
(e)   Occurrence and distribution of fractures
(f)   Characteristics and distribution of voids
(g)   Presence of contaminating substances
(h)   Evidence of cement-aggregate reaction
(i)   Proportion of unhydrated granules of cement
(j)   Presence of mineral admixtures
(k)   Volumetric proportions of aggregate, cement paste, and air voids
(l)   Air content and various parameters of the air void system.

Care should be taken in selecting core samples to ensure a maximum true representation of the material and to avoid unusually poor or unsound materials.

Sampling should include near-surface concrete as well as concrete at depth because of the possible affect of depth on the properties of the materials and the defects. Also, the number and size of samples should be carefully selected to permit all necessary laboratory tests. It is important to use virgin samples in order that there be no influence from the previous tests.

Evaluation should be performed only by a qualified petrographer who is familiar with problems commonly encountered with concrete. The petrographer should be consulted before proceeding with the core sampling and should be provided with as much preconstruction, construction, and post-construction performance information as possible to assist in the evaluation. Following petrographic examination and analysis, the petrographer prepares a report summarizing observations and conclusions regarding the suitability of the aggregate under the anticipated conditions of service and indicating any possible effects the aggregate may have had on problems encountered with the concrete.

ASTM C 295 "Recommended Practice for Petrographic Examination of Aggregate for Concrete," and ASTM C 856 "Recommended Practice of Petrographic Examination of Hardened Concrete" give more complete descriptions of this test.

PROBLEMS WITH CONCRETE–A Historic Building Checklist

When investigating weathering failures and deterioration of concrete it is important to remember that they are the result of a combination of factors.

- Analysis is crucial before repair or restoration work can proceed.
- Remember that concrete requires exceptionally strict quality control in both design and placing.
- Criteria governing the correct ingredients and building code requirements governing the use of concrete have developed over a long period of time.
- —What is perceived as inadequate design today was possibly originally in accordance with existing codes and practice of the time. As an example the concrete cover over reinforcing specified at the turn of the century was substantially less than is required today.
- A critical factor in evaluating concrete is the composition of the original mix.
- Characteristics of old and new concrete must be compatible if failure is to be avoided.
- —Concrete mixes are carefully specified today which prescribe binders as well as artificial aggregates.
- —Construction practice of the past did not have as stringent requirements. As a result some of the observed deterioration may be due to these causes.
- The use of natural cements such as Rosedale was common.
- —These cements manufactured from natural rock have an uneven composition and therefore their quality varies widely.
- —Natural cements remained in use even after more carefully controlled artificial cements such as Portland were in use.
- White Portland cement was quite costly before WW II.

Wooden forms used as wall finish in a New Haven School. (Photo by F. Wilson)

—As a result it was not manufactured domestically. Lime was often used where light-colored concrete was desired.

• Today the composition and quality of aggregates is closely controlled.

—In the past the problems of aggregate shrinkage, ferrosulfide contamination, and alkali-aggregate reactions were not always recognized.

—Chemical reactions between aggregates and sodium or potassium alkalis formed during cement hydration were poorly understood.

—The contribution of grading and size distribution of aggregates to shrinkage cracking or the uneven coverage of embedded reinforcement was not as carefully considered as it is today.

• The contamination of aggregates by sodium chloride as a result of using beach sand was not always carefully considered.

• Crushed limestone was commonly used as aggregate when a natural stonelike cast stone appearance was desired. The erosion and deterioration of the limestone was accepted as an integral part of the weathering of the concrete.

• Excessive chloride accumulation results in surface efflorescence.

—This is particularly true in climates with extensive wet-dry cycles.

—The resulting salt crystals damage concrete surfaces and cause them to powder and disintegrate.

• The use of concrete additives is not new, although they were not always used as extensively prior to the development of the petrochemical industry.

—Previously mineral pigments were used without full comprehension of their effects on the concrete's strength.
—Sterate to obtain waterproofing or chlorides for rapid set were known techniques.
—There did not seem to be sufficient understanding of the influence of calcium or sodium chlorides.
• The most serious problem encountered in concrete structures is the corrosion of the embedded reinforcement or structural steel.
—Normally this is inhibited by high alkalinity of uncarboned concrete. If alkaline conditions disappear, the steel becomes susceptible to corrosion. Alkalinity may be reduced as a normal weathering function. Continuous exposure to outside air results in a slow but gradual carbonation. The formation of carbonates produced from water and carbon dioxide present in air causes the lowering of alkalinity. If concrete cover is sufficient this process occurs very slowly over a long period of time and does not markedly affect the condition of the reinforcement.
—Thin or permeable cover may cause a serious problem. Concrete with poorly graded aggregates or a high water-cement ratio, which is likely to be porous, is very susceptible to carbonation.
• Minor defects in concrete surface such as shrinkage cracking accelerate deterioration.
• Sufficiently large quantities of calcium chlorides promote corrosion despite high alkaline conditions.

Visual observations

• Rust-staining and cracking of the exterior surfaces are generally indicative of corrosion of reinforcement.
• Regular crack patterns following the direction of reinforcement or the stirrups are generally a strong indication of this condition.
—If corrosion has progressed dangerously the concrete cover will have spalled, completely exposing rebars.
—If the rebars have seriously corroded the effective cross section of sound steel may no longer be sufficient, resulting in structural failure.
Serious but more difficult to establish are deterioration problems inherent in the design and detailing of the concrete structure itself.

• Location and size of reinforcement may vary widely.
—Rebars as known today, evolved about 1900. They previously had a variety of shapes and yield strengths. Reinforcement configurations have changed over time.

Thermal exposure may have a direct impact on structural performance and durability.

• Expansion joints were not always included in design detailing.

—The structure became vulnerable to extensive cracking sometimes buckling of sections. Thermal cracks generally appear at corners and ends of walls, unlike cracking caused by corrosion of reinforcement, which is generally evenly patterned over large areas.

Design of exposed units or portions may have a direct relationship to weathering and deterioration.

• Too-thin face covering for cast stone may result in extensive delamination.
—Cast stone window lintels with concrete covers considered adequate for vertical wall sections may exhibit marks of deterioration due to excess water over sills or belt courses causing saturation and subsequent reinforcement corrosion.

Rarely is the deterioration of a concrete structure the direct result of structural failures.

• Particularly when the building has stood for a period of time.
• Subsequent changes in loading or use patterns may cause deterioration to develop.
—Overloading may have caused deflection, cracking.

PRECAST CONCRETE PANELS

From the early 1960s on architects began to use precast concrete panels in place of the ubiquitous metal curtain wall so popular during the 1950s.

Precasting exposed aggregate panels became a special branch of concrete manufacture. It was, and remains, a particularly skilled activity having little relation or technical similarity to its historic "cast stone" predecessor. The following description of problems and pitfalls or precasting exposed aggregate panels, by Ray McNeal and A. J. Tenzer, is a valuable guide to diagnosing difficulties in these panels now that they have experienced thermal loads environmental hazards, and the freeze thaw cycle for several decades.

Exposed Aggregates—A Precasting Primer

Exposed-aggregate concrete panels, with their variety of colors, designs, and textures, can be counted among architecture's most versatile developments. They also create many problems in construction and disappointments in service, but this fact is less surprising than its corollary: sometimes they do not.

What makes a totally successful installation of precast exposed aggregate so surprising is that it can only result from the most delicate coordination of design, engineering, specification, manufacturing, and erection. Problems arising at each stage of manufacture require an understanding of what went before and what will follow after; sometimes, the information on which judgments must be based is neither unavailable or already buried in a batch of cured concrete.

Precast concrete panels, New York City (Photo by F. Wilson)

Although causes may vary, failures can usually be traced to the Portland cement concrete that is used in most exposed-aggregate precast panels. Weathering and stress cracks, spalling, or disintegration are most common and most dangerous, because they permit moisture penetration that causes progressive deterioration.

The binder may fail to hold all of the exposed aggregate, resulting in unattractive voids that also permit progressive damage. Staining and discoloration are separate problems. They may or may not be related to a structural weakness.

In consideration of the structure of a typical panel and how it is affected by design, specification, quality control and manufacturing, we may find where troubles begin.

The exposed face of a precast panel is rarely "ordinary" concrete. For the sake of appearance and texture, it is usually made of selected, tightly graded—and frequently costly—aggregate mixed with clean white sand and cement, and more often than not with pigment or coloring agent. These expensive ingredients are not needed in the unexposed, backup portion (some are undesirable, since they weaken concrete), but for strength and durability each panel has to be a monolithic mass.

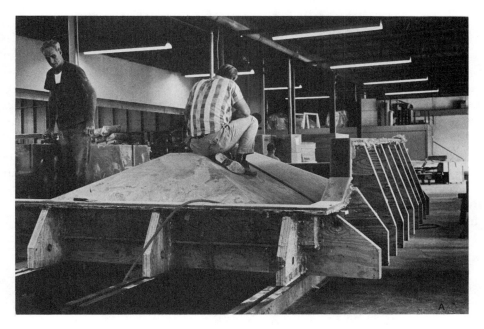

Fiberglass form in precasting shop. (Photo by F. Wilson)

Therefore, to keep costs down and strength up, most panels are cast in two steps. First the reinforced backup is cast with a mix formulated without regard to appearance. Then before it can take an initial set, separately mixed concrete for the face layer is cast over it. The two layers are consolidated, usually by mechanical vibration.

Stacking precast forms in concrete yard. (Photo by F. Wilson)

The panel then has to be cured. Steam curing is commonly used because it yields an ultimate strength nearly as good as water cure, and is a great deal faster and therefore less expensive. The heat of the steam chamber speeds the initial curing action of the concrete, and the moisture-saturated environment prevents loss of water that the concrete needs for a full, high-strength cure. The panel must be kept moist for at least seven days after it is removed from the steam. This can be done either by placing it in a high-humidity chamber or by covering it with a suitable plastic membrane.

The final manufacturing step consists of exposing the surface aggregate. In one method it is etched with solutions of hydrochloric acid. In another the curing time is retarded by applying chemicals to the surface of the concrete mix during casting. The shallow, uncured layer can be scrubbed away after the bulk of the mix has set hard enough to be unaffected by hosing and brushing.

Mistakes Multiply

As simple as these procedures sound, they are full of pitfalls. Precast panel problems can begin at any stage, from design onward; the earlier they start, the more chances they have to multiply along the way. And good components and systems may fail because they are incompatible with each other.

The initial design and specification of exposed-aggregate precast panels is partly architectural, partly an engineering function. Requirements of size, thickness, color, texture, strength, loading and resistance to probable environments have to be translated into concrete—component ratios, density, air-entrainment, pigments, aggregate type, size and grading, and the size and placement of reinforcement. All of these are important to a sound and durable end product; most are critical.

The specification of concrete involves many compromises and trade-offs. Assuming that all components are of the quality and purity called for (they are not always), the design must still balance cost and appearance against tensile, flexural, and compressive strengths, density, absorption, and similar physical factors. Most of these are determined by the basic formulation—that is, the ratio of cement to total aggregate to water. Subsidiary relationships, such as fine to coarse aggregate, exposed aggregate mix to back-up mix, and all of these to pigment, air-entraining agents, and chemical accelerators or retarders, if any, are almost equally important.

Where strength is a factor, high density is, of course, desirable. Exposed-aggregate panels are rarely loadbearing; but they have to be strong enough to withstand strains of handling and erection, settling and possible misalignment, and wind loading. Higher density also gives better resistance to weathering, disintegration, water penetration, dirt retention, and usually loss of exposed aggregate.

Since it is not usually practical to compound exposed concrete for total resistance to moisture penetration, the surface layer should entrain between 4 and 7% air. This makes it less vulnerable to freeze damage from any water that does get in. Since air-entrainment reduces strength, the loss should be compensated for in mix design.

The same is true of pigment. It contributes nothing but color to the mortar and should therefore be calculated as part of the inert filler. Too often, it is simply added to the total without regard for the fact that it can change the overall proportion in the direction of lower strength and poorer bonding. Some types may also react chemically with some types of cement, resulting in staining, uneven color, or early disintegration. Impurities in pigments can have similar or worse effects.

The Stone's the Thing

To a designer of exposed-aggregate panels, the component of greatest interest, quite naturally, is the exposed aggregate. Frequently, it is also the most troublesome. Although appealing to the eye, some types of aggregate are basically unsuited for this use because they react with the alkali in cement; in exposed faces, they tend to expand and pop out of the mortar. Some lightweight aggregates are good to look at and would seem to be desirable for minimizing panel weight. However, they should usually be avoided, because they are apt to absorb moisture which, in exposed surfaces, would promote cracking and deterioration. In time, they would probably crumble out and damage the appearance as well as the strength of the panel.

When specifying aggregate, an architect should also consider how the surface is to be exposed. If acid is to be used, it is important that the aggregate be acid resistant. Otherwise the color (and possibly the strength) for which the stone is selected may be leached out before the panel ever gets on a wall. Even if chemical retarders are to be used for exposing the surface, it is best to specify acid-resistant aggregate, for, in steam curing, the surface mortar sometimes sets up too quickly for successful wash exposure. When this happens, acid has to be used, regardless of the manufacturing plan. The size of coarse aggregate is as important to durability as it is to appearance. Even more important is the graduation of sizes, for whatever the design reasons for mixing small and large aggregate in an exposed face, they are more than offset by disadvantages in function. Inevitably, if the surface is etched enough to give proper exposure to large stones, small ones will scale out sooner or later—probably sooner.

Aggregate size, in the base mix as well as the surface mix, must also be considered in relation to panel thickness. As a rule of thumb, the largest piece should be no more than one-quarter the thickness of the panel.

Size may also be governed by the need to keep reinforcing steel at least ¾ in. below the surface. In exposed-aggregate panels, this measurement should be taken from the most deeply etched valley. Some water penetration has to be expected, and, if any moisture does reach the steel, the resulting oxidation and expansion will surely cause progressive damage. So, if a safe depth cannot be maintained for reasons of design or structure, specify galvanized reinforcement and coat the ends of cut bars with a zinc chromate primer.

No Single Cause

Careful, knowledgeable specification is essential to a satisfactory end product, but it is not enough. Everything an architect or engineer plans for an exposed aggregate panel can be undone by poor manufacturing techniques, or every by good techniques that are imperfectly executed.

Good manufacturing must begin with careful inspection, testing, and quality control of every batch of component materials. All aggregate—coarse, fine, exposed, and backup—should be tested for soundness, alkaline reaction, chloride content, cleanliness, and freedom from deleterious substances and minerals. They should also be checked for size and graduations of sizes. Substandard aggregates or mixed-in impurities, as has been noted, create time-delayed problems by expanding, crumbling, reacting chemically, staining, blooming, or just weakening the concrete mass.

Batching and proportioning can also be critical. In the inspection of problem jobs, it is routine to check core samples for physical properties. Probably the most common finding is that compressive strength is far below design specification. This is probably most often due to nonuniform batching and poor placement of mixed batches in the forms. There is a natural tendency for coarser aggregate to settle to the bottom. Unless this is checked by careful batching, mixing, and slump control, it will leave strength-robbing voids in the cured concrete.

Many other factors can contribute to nonuniform, subspecification concrete. They are explained below, along with preventive measures.

Water-Content Variations. Even when water is carefully measured, calculations can be upset by rain or air moisture settling on the surface of stored aggregate. This must be allowed for; it should be recalculated each time the batching storage bin is filled from the stockpile. Moisture varies as the overnight accumulation dries out during the day and as the scoop reaches the inner, drier part of the pile. This kind of sliding-scale calculation may not defeat an engineer, but it could pose problems for the mechanic on the stockpile bucket. Such problems could be avoided, or at least minimized, by requiring that all aggregates be stored under cover.

Cement. Cement quality is not necessarily uniform. Each incoming supply batch should be tested not only to confirm freshness and quality, but also to make sure of batch-to-batch uniformity. Even slight variations in color, setting time, or processing requirements between batches mixed together can cause unpredictable results. Quality control at this point should also insure that none of the cement used is already partly hydrated.

Consolidation. The separate pours of surface and backup concrete have to be mechanically tied together at the interface. This is done by vibrating the combined mass before either layer has obtained an initial set. Unless the process is carefully handled, it could bleed some of the backup mortar to the surface. This would be no problem unless the mortars were of a different color. But they frequently are. Somewhat worse, it could settle large aggregate to the bottom and weaken the concrete. Or it could cluster or rearrange stones that should be spread out for visual effect.

Curing and Handling. This should be routine, but not infrequently it is routinely bad. For a high-strength cure, concrete has to be kept thoroughly and continuously moist for at least seven days, preferably 14. Panels should be processed at a uniform rate. It is best not to handle them at all until they have taken a good set, and then they should be handled with great care through the remaining cure cycle. Premature or rough movement can set up internal cracks or strains from which concrete can never recover. The dangers are greatest when panels are moved through cure stages, and when they are removed from their forms. Perhaps some banging is necessary at this point, but in a good shop, workmen will not seize upon form release as an excuse for letting out their aggressions.

Exposing Surface Aggregate. Whether exposure is accomplished by acid etching or by the retardation method (or by both, if steam curing hardens the surface too fast for successful washout), it is important that the treatment not remove too much of the binder. Acids have to be applied carefully in proper strengths for controlled times or they will penetrate unevenly and too deeply. This will result in an inert binder that obviously cannot hold the aggregate, and dead cement can wash out later and invite water penetration with all the resultant damage of steel oxidation and freeze-thaw damage. Chemical retarders must be handled with equal care to insure adequate, but not excessive, binder washout.

Testing Panels for Soundness. If there are going to be any problems, it is clearly better to know about them before panels are hung on a structure. In a good quality control procedure, test cylinders and slabs will be taken from each mix batch from which panels are cast. Before any panel is used, its matching samples should be cured for 28 days and then thoroughly tested for tensile and compressive strength, absorption, freeze-thaw stability, and other relevant properties. And, of course, the samples should be closely examined for hairline cracks, voids, discoloration, or other flaws.

Haste Makes Waste

Care at every stage of design, manufacture, and testing can eliminate most problems of exposed-aggregate panels. But until men become infallible, they will not eliminate all. Further checks of the finished structure should therefore be made both early and periodically. For every detection and treatment of minor faults can sometimes keep larger ones from developing.

If inspection does reveal problems, it then becomes important to analyze their causes and forecast probable effects. Minor loss of aggregate may not be serious, but if stiff brushing removes a significant amount of stone and powders out the matrix, it is likely that weathering effects will be severe.

Cracks, too, have to be carefully analyzed. If there are many of them, it may be necessary to remove a panel and project its weatherability by accelerated freeze-thaw cycling. Minor hairline cracks, if there are not too many of them, can probably be checked by a sprayed or brushed-in water repellent. This should not be attempted, however, until the future course of crack development has been projected as accurately as possible. Many surface waterproofers are based on silicones that repel not only water, but most other materials as well. If cracks should open up, the silicone treatment will not only be ineffective, but will prevent the adhesion of new grout or slurry that could repair the damage. Therefore, repellent surface treatment should not be used unless there is good reason for believing that serious cracks will not develop for at least three to five years. After that the silicones will be sufficiently worn away to permit further sealing. And, in the meantine, they should afford considerable protection for both the concrete and the building.

Serious and progressive cracking presents other problems that, through exposure of reinforcing steel, could affect structural soundness. Problems like these will probably require rebuilding of the matrix with carefully worked-in grouting compounds.

Polymer types of emulsion solutions have been successfully used to overcome binder and stone defects. Corrective compounds must be developed and tailored very carefully, so that their physical properties and coefficients of contraction and expansion will be similar to those of the original panels.

REFERENCE

Tenzer, A. J., and McNeal, Ray, of the Building Materials Research Institute, Inc., N.Y.C., *P/A,* Oct. 1966, p. 221.

VIII. Cleaning Building Surfaces

Definition—Dirt

Cleaning Masonry

What is Dirt?

Considerations of Building Construction

Types of Cleaning

Environmental Concerns

Potential Cleaning Problems

Water Cleaning

Chemical Cleaning

Mechanical Cleaning

Problems with Water-Repellent and Waterproof Coatings

How Building Materials React to Abrasive Cleaning Methods
Brick and Architectural Terra- Cotta
Plaster and Stucco
Building Stones
Wood
Metals

Gentlest Means Possible

Mitigating the Effects of Abrasive Cleaning

Visual Inspection of Painted Surfaces

DEFINITION—Dirt

Dirt can be defined as a surface deposit of finely divided solids, generally held together by small amounts of organic material. The solids consist primarily of carbon soot, siliceous dust, and other airborne particulates such as inorganic sulfates. The nature of the organic binder that holds these fine particles together is chemically more difficult to characterize. It is probably largely made up of aliphatic hydrocarbons resulting from the incomplete combustion of both wood and fossil fuels.

Dirt adheres to masonry surfaces by a combination of absorption and electrostatic attraction. Under such circumstances, the persistent dirtiness of any localized area of a building or monument is determined by a number of complex parameters, including surface texture and physical orientation (with regard to such directional effects as wind, rain, and water run-off).

At times, dirt layers are held strongly in place by partial incorporation in a crust consisting of deterioration products, efflorescences, recrystalized carbonate

Dirt on the escutcheon, lower Manhattan. (Photo by F. Wilson)

minerals and leached cementing materials. Such accretions may be significantly resistant to removal by simple methods.

A suitable cleaning method depends on the composition of the dirt itself, and the surface of the material on which it rests. The real challenge is the extraordinary variety of geological and man-made materials used in masonry construction, and in the subtle differences of the long-term effects of the immediate environment in each individual case.

CLEANING MASONRY

A strong argument to support masonry cleaning is that it can assist in the prevention of future deterioration. Dirt promotes chemical reactions that induce the absorption of atmospheric gases into any moisture that may be present. Dirty areas remain wet for a longer period of time after rainfall, accelerating the reactivity of the masonry to common pollutants.

Increased residual dampness can amplify deterioration such as that associated with freeze-thaw cycling. When a relatively high surface moisture content is maintained, microvegetation can thrive. Biological accretions stimulate surface disintegration, dissolution, and staining.

However, the cleaning of masonry structures is not always an appropriate step. The use of improper materials or techniques in the cleaning process can cause considerable damage. Some combinations of physical condition and chemical reactivity may rule out cleaning entirely. It is conceivable that the presence of certain types of dirm films (especially those with a very high organic content) might be beneficial, and therefore worth maintaining intact. Caution is essential.

What is Dirt?

The general nature and source of building dirt must be determined to remove it in the most effective, least harmful manner. Soot and smoke, for example, require a different method of cleaning than oil stains or bird droppings. The "dirt" may be weathered or discolored portions of the masonry itself. Common cleaning problems include metal stains, rust or copper stains, and organic matter such as the tendrils left on the masonry after removal of ivy. The source of dirt, such as coal soot, may no longer be a factor in planning for longer-term maintenance, or it may be a continuing source of problems.

Considerations of Building Construction

Incorrectly chosen cleaning products can cause damaging chemical reactions. The effect of acidic cleaners on marble and limestone generally is recognized. Other masonry products also are subject to adverse chemical reactions with incompatible cleaning products.

Some chemicals may have a corrosive effect on paint or glass. The portions of building elements most vulnerable to deterioration may not be visible, for example, embedded ends of iron window bars. Other totally unseen items, such as iron cramps or ties which hold the masonry to the structural frame will be subject to corrosion from the use of chemicals or even from plain water.

Earlier waterproofing applications may make cleaning difficult. Repairs may have been stained to match the building, and cleaning may make these differences apparent. Salts or other snow removal chemicals used near the building may have dissolved and been absorbed into the masonry, causing potentially serious problems of spalling or efflorescence.

Types of Cleaning

Cleaning methods are divided generally into three major groups; water, chemical, and mechanical (abrasive). Water softens the dirt and rinses the deposits from the surface.

Chemical cleaners react with dirt and masonry or other material to hasten the removal process. The deposits and excess chemicals then are rinsed away with water. Mechanical methods include grit blasting (usually sand blasting), grinders, and sanding discs, which remove the dirt by abrasion usually are followed by a water rinse.

Environmental Concerns

Chemical cleaners, even though diluted, may damage trees, shrubs, grass and plants. Animal life, ranging from domestic pets to songbirds to earthworms, can be affected by the runoff. Mechanical methods can produce hazards through the creation of airborne dust.

Cleaning may also cause property damage. Wind drift may carry cleaning chemicals onto nearby automobiles, causing etching of the glass or spotting of the paint finish. Similarly, airborne dust can enter surrounding buildings, and excess water can collect in nearby yards and basements.

Personal Safety. Both acidic and alkaline chemical cleaners can cause serious injury to cleaning operators and passerby; injuries can be caused by chemicals in both liquid and vapor forms. Mechanical methods cause dust, posing serious health hazards, particularly if they contain silica. Steam cleaning is dangerous because of high temperatures.

Tests. Tests should be applied to an area of sufficient size to give a true indication of effectiveness. The test patch should include at least a square yard, and with large stones, include several stones and motar joints. It should be remembered that a single building may have several types of masonry materials and similar materials may have different surface finishes; each of these differing areas should be tested separately. The results of tests may indicate the several methods of cleaning should be used on a single building.

When feasible, test areas should be allowed to weather for an extended period prior to evaluation. A waiting period of a full year exposes the masonry to a full range of seasons.

POTENTIAL CLEANING PROBLEMS

Water Cleaning

Water cleaning methods include: (1) low-pressure wash over an extended period, (2) moderate- to high-pressure wash, and (3) steam. Bristle brushes frequently are used to supplement the water wash. All joints, including mortar and sealants, must be sound to minimize water penetration to the interior.

Porous masonry may absorb water and cause damage within the wall or on interior surfaces. Normally water pentrates only partially even moderately absorbent masonry materials.

Excess water may also bring soluble salts to the surface, forming efflorescences. In dry climates water may evaporate inside the masonry, leaving salts in back of the surface. (However, efflorescence can usually be traced to other sources.) Chemicals such as iron and copper are present in the water supply; even "soft" water may contain harmful amounts of these chemicals. Water cleaning cannot be used during cold weather. Water within the masonry can freeze, causing spalling and cracking. Since a wall may take over a week to dry after cleaning, no water cleaning should be permitted for several days prior to the first average frost date, or even earlier if local forecasts predict cold weather.

In spite of these potential problems, water methods generally are the simplest, the safest for building and environment, and the least expensive.

Chemical Cleaning

Since most chemical cleaners are water based, they have many of the potential problems of plain water. Chemical cleaners have other problems as well. Some types of masonry are subject to direct attack by cleaning chemicals. Marble and limestone are dissolved easily by acidic cleaners, even in diluted form. Another problem may be a chemically induced change in the color of the masonry. The cleaner also may leave a hazy residue in spite of heavy rinsing. In addition, chemicals can react with components of mortar, stone, or brick to create soluble salts, forming efflorescences. Historic brick buildings are particularly susceptible to damage from hydrocholoric (muriatic) acid.

Mechanical Cleaning

Grit blasters, grinders, and sanding disks operate by abrading the dirt off the surface of the masonry. Since the abrasives do not differentiate between dirt and masonry, some erosion of the masonry surface is inevitable with mechanical methods, especially blasting. Although a skilled operator can minimize this erosion, it will still take place. In the case of brick, soft stone, detailed carvings, or polished surfaces, even minimal erosion may be harmful; brick, a fired product, is hardest on the outside where the temperatures were highest. The loss of this skin of the brick exposes the softer, inner portion to more rapid deterioration.

Abrasion of intricate details causes a rounding of sharp corners and other loss of delicate features, while abrasion of polished surfaces removes the smooth stone surface.

Blasting, in most cases, leaves minute pits on masonry surfaces. This additional roughness increases the area surface on which dirt can settle and on which pollutants can react.

Mortar joints, especially those with lime mortar, can be eroded by mechanical cleaning. Joints constitute a significant portion of the masonry surface—as much as 20% in a brick wall. The erosion of the mortar joint may permit increased water penetration.

Problems with Water-Repellent and Waterproof Coatings

Water penetration to the interior is usually not caused by porous masonry but by deteriorated gutters, downspouts, or mortar, by capillary moisture from the ground (rising damp), or by condensation. Coatings will not solve these problems. In the case of rising damp, the coatings will allow the water to go even higher because of the retarded rate of evaporation. Claim is made that coatings keep dirt and pollutants from collecting on building surfaces, reducing the requirements for future cleaning. While this at times may be true, at others coatings actually retain the dirt more than uncoated masonry. Coatings can cause greater deterioration of the masonry than that caused by the pollution.

Masonry coatings are of two types: waterproof and water repellent. *Waterproof coatings* seal the surface from liquid water and from water vapor; they usually are opaque, such as bituminous coatings and some paints. *Water repellents* keep liquid water from penetrating the surface but allow water vapor to enter and leave through the pores of the masonry. They are usually transparent, such as silicone coatings, although they may change the reflective quality and thus the appearance of the masonry.

Water-repellent coatings that do not seal the surface against water vapor entering the wall may cause problems. Once inside vapor may condense at cold spots, resulting in a liquid. Water within a wall can do damage. Soluble salts can do further damage. Salts are frequently present in masonry, either from the mortar or from the masonry units themselves. Liquid water can dissolve these salts and transport them toward the surface, where efflorescences appear upon evaporation. These are unsightly but usually easily removed often washed away by the rain.

The presence of a water-repellent coating, however, hampers water and dissolved salts from rising to the surface completely. The salts are deposited slightly below the surface of the masonry as the water evaporates through the pores. Over time, salt crystals grow and develop substantial pressures which will spall the masonry, detaching it to the depth of crystal growth. Such buildups may require several years.

Test patches for coating generally do not allow adequate evaluation of the treatment, because water may enter and leave through the surrounding untreated areas, thus flushing away the salt build-up. In addition, salt deposits may not cause visible damage until well after the patch has been evaluated.

There are, however, uses for water repellents and waterproofings. Sandblasted brick, for example, may become so porous that paint or other type of coating is essential. In other instances, damage caused by local pollution may be greater than the potential damage from coatings. Generally, coatings are not necessary unless there is a specific problem which they will solve. If the problem occurs on only a portion of the masonry, it probably is best to treat only the problem area rather than the entire building. Extreme exposures such as parapets, for example, or portions of the building subject to driving rains can be treated more effectively and less expensively than the entire building.

How Building Materials React to Abrasive Cleaning Methods

Brick and Archtectural Terra Cotta

When these materials are cleaned abrasively, the hard, outer layer, closest to the heat of the kiln, is eroded, leaving the soft, inner core exposed and susceptible to accelerated weathering. Glazed architectural terra cotta and ceramic veneer have a baked-on glaze which is also easily damaged by abrasive cleaning. Glazed architectural terra cotta was designed for easy maintenance, and generally can be cleaned using detergent and water; but chemicals or steam may be needed to remove more persistent stains. Large areas of brick or architectural terra cotta which have been painted are best left painted, or repainted if necessary.

Plaster and Stucco

Plaster and stucco are softer than brick or terra cotta. If treated abrasively they will simply disintegrate. Plaster and stucco are usually removed from underlying surfaces by abrasive techniques.

Building Stones

Building stones are cut from the three main categories of natural rock: dense, igneous rock such as granite; sandy, sedimentary rock such as limestone; and crystalline, metamorphic rock such as marble. Unlike kiln-dried masonry materials such as brick and architectural terra cotta, building stones are generally homogeneous in character at the time of a building's construction. However, as the stone is exposed to weathering and environmental pollutants, the surface may become friable, or develop a protective skin or patina. These outer surfaces are susceptible to damage by abrasion or improper chemical cleaning.

Building stones are frequently cut into ashlar blocks or dressed with tool marks that give the building surface a specific texture and contribute to its character. Such detailing is easily damaged by abrasive cleaning techniques; the pattern of tooling or cutting is erased, and the crisp lines of moldings or carvings are worn or pitted.

Occasionally, it may be possible to clean small areas of rough-cut granite, limestone, or sandstone suffering heavy dirt encrustation by using the wet grit method, in which a small amount of abrasive material is injected into a controlled, pressurized water stream. However, this technique requires very careful supervision in order to prevent damage to the stone. Polished or honed marble or granite should never be treated abrasively, as the abrasion removes the finish in much the way glass would be etched or "frosted" by such a process. It is generally preferable to underclean, as too strong a cleaning procedure will erode the stone, exposing a new and increased surface area to collect atmospheric moisture and dirt. Removing paint, stains or graffitti from most types of stone may be accomplished by a chemical treatment carefully selected to best handle the removal of the particular type of paint or stain without damaging the stone.

Wood

Most types of wood used for building are soft, fibrous, and porous, and are thus particularly susceptible to damage by abrasive cleaning. Because summer wood is softer it will be worn away by abrasive blasting or power tools, leaving an uneven surface with the grain raised and frayed. Once this has occurred, it is almost impossible to achieve a smooth surface again except by extensive sanding, which is expensive and will quickly negate any costs saved earlier by sandblasting. Such harsh cleaning treatment also obliterates historic tool marks, fine carving, and detailing, which precludes its use on any interior or exterior woodwork which has been hand planed, milled, or carved.

Metals

Like stone, metals are another group of building materials which vary considerably in hardness and durability. Softer metals which are used architecturally, such as tin, zinc, lead, copper, or aluminum, generally should not be cleaned abrasively as the process deforms and destroys the original surface texture and appearance, as well as the acquired patina. Much applied architectural metalwork used on historic buildings (tin, zinc, lead, and copper) is often quite thin and soft, and therefore susceptible to denting and pitting. Galvanized sheet metal is especially vulnerable, as abrasive treatment would wear away the protective galvanized layer.

In the late 19th and early 20th Centuries, these metals were often cut, pressed or otherwise shaped from sheets of metal into a wide variety of practical uses such as roofs, gutters, and flashing, and façde ornamentation such as cornices, friezes, dormers, panels, cupolas, oriel windows, etc. The architecture of the 1920s and 1930s made use of metals such as chrome, nickel alloys, aluminum, and stainless steel in decorative exterior panels, window frames, and doorways. Harsh abrasive blasting would destroy the original surface finish of most of these metals, and increase the possibility of corrosion.

However, conservation specialists are now employing a sensitive technique of glass bead peening to clean some of the harder metals, in particular large bronze outdoor sculpture. Some fine (75–125 micron) glass beads are used at a low pressure of 60–80 psi. Because these glass beads are completely spherical, there are no sharp edges to cut the surface of the metal. After cleaning, these statues undergo a lengthy process of polishing. Coatings are applied which protect the surface from corrosion, but they must be renewed every 3–5 years. A similar delicate cleaning technique employing glass beads has been used in Europe to clean historic masonry structures without causing damage, but at this time the process has not been tested sufficiently in the United States.

Sometimes a very fine, smooth sand is used at a low pressure to clean or remove paint and corrosion from copper flashing and other metal building components. Restoration architects have recently found that a mixture of crushed walnut shells and copper slag at a pressure of approximately 200 psi was the only way to remove corrosion successfully from a mid-19th Century terne-coated iron roof. Metal cleaned in this manner must be painted immediately to prevent rapid recurrence of corrosion. It is thought that these methods, work harden the surface by compressing the outer layer, and actually may be good for the surface of the metal. But the extremely complex nature and the time required by such processes make it very expensive and impractical for large-scale use at this time.

Cast and wrought iron architectural elements may be gently sandblasted or abrasively cleaned using a wire brush to remove layers of paint, rust, and corrosion. Sandblasting was, in fact, developed originally as an efficient maintenance procedure for engineering and industrial structures and heavy machinery—iron and steel bridges, machine tool frames, engine frames, and railroad rolling stock—in order to clean and prepare them for repainting. Because iron is hard, with a naturally somewhat uneven surface, it will not be noticeably damaged by controlled abrasion. Such treatment will result in a small amount of pitting, but this pitted surface creates a good surface for paint. The iron must be repainted immediately to prevent corrosion. Any abrasive cleaning of metal building components will also remove the caulking from joints and around openings. Such areas must be recaulked quickly to prevent moisture from entering and rusting the metal, or causing deterioration of building elements inside the structure.

Gentlest Means Possible

There are alternative means of removing dirt, stains, and paint from building surfaces that can be recommended as more efficient and less destructive than abrasive techniques. The gentlest means possible of removing dirt from a building surface can be achieved by using a low-pressure water wash, scrubbing areas of more persistent grime with a natural bristle (never metal) brush. Steam cleaning can also be used effectively to clean some historic building fabric. Low-pressure water or steam will soften the dirt and cause the deposits to rise to the surface to be washed away.

A third cleaning technique which may be recommended to remove dirt, as well as stains, graffiti or paint, involves the use of commercially available chemical cleaners or paint removers which, when applied to masonry, loosen or dissolve the dirt or stains. These cleaning agents may be used in combination with water or steam, followed by a clear water wash to remove the residue of dirt and the chemical cleaners from the masonry. A natural bristle brush may also facilitate this type of chemically assisted cleaning, particularly in areas of heavy dirt deposits or stains, and a wooden scraper can be useful in removing thick encrustations of soot. A limewash or absorbent talc, whiting, or clay poultice with a solvent can be used effectively to draw out salts or stains from the surface of the selected areas of a building facade. It is almost impossible to remove paint from masonry surfaces without causing some damage to the masonry, and it is best to leave the surfaces as they are or repaint them if necessary.

Some physicists are experimenting with the use of pulsed laser beams and xenon flash lamps for cleaning historic masonry surfaces. At this time it is a slow, expensive cleaning method, but its initial success indicates that it may have an increasingly important role in the future.

There are many chemical paint removers which, when applied to painted wood, soften and dissolve the paint so that it can be scraped off by hand. Peeling paint can be removed from wood by hand scraping and sanding. Particularly thick layers of paint can be softened with a heat gun or heat plate, providing appropriate precautions are taken, and the paint film scraped off by hand. Too much heat applied to the same spot can burn the wood, and the fumes caused by burning paint are dangerous to inhale, and can be explosive. Furthermore, the hot air from heat guns can start fires in the building cavity. Thus, adequate ventilation is important when using a heat gun or heat plate, as well as when using a chemical stripper. A torch or open flame should never be used.

Each kind of masonry has a unique composition and reacts differently with various chemical cleaning substances. Water and chemicals may interact with minerals in stone and cause new types of stains to leach out to the surface immediately, or more gradually in a delayed reaction. What may be a safe and effective cleaner for certain stain on one type of stone, may leave unattractive discolorations on another stone, or totally dissolve a third type.

Mitigating the Effects of Abrasive Cleaning

Certain methods can be employed to preserve a historic building exterior that has been damaged by abrasive methods. Wood that has been sandblasted will exhibit a frayed or "fuzzed" surface, or a harder wood will have an exaggerated raised grain. The only way to remove this rough surface or to smooth the grain is by sanding. Sandblasted wood will weather faster and present a continuing and ever worsening maintenance problem. Such wood, after sanding, should be painted or given a clear surface coating to protect the wood and allow for somewhat easier maintenance.

There are few successful preservation treatments that may be applied to grit-blasted exterior masonry. Harder, denser stone may have suffered only a loss of crisp edges or tool marks, or other indications of craft technique. If the stone has a compact and uniform composition, it should continue to weather with little additional deterioration. But some types of sandstone, marble, and limestone will weather at an accelerated rate once their protective quarry crust or patina has been removed.

Softer types of masonry, particularly brick and architectural terra cotta, are the most likely to require some remedial treatment if they have been abrasively cleaned. Old brick, being essentially a soft, baked clay product, is more susceptible to increasing deterioration when its hard, outer skin is removed through abrasive techniques. This problem can be minimized by painting the brick. An alternative is to treat it with clear sealer or surface coating, but this will give the masonry a glossy or shiny look. It is usually preferable to paint rather than apply a transparent sealer. If a brick surface has been so extensively damaged by abrasive cleanig and weathering that spalling has begun, it may be necessary to cover the walls with stucco, if it will adhere.

Of course, the application of paint, a clear surface coating (sealer) or stucco to deteriorating masonry means that the historical appearance will be sacrificed in an attempt to conserve the historic building materials. At this point it is more important to try to preserve the brick, and there is little choice but to protect it from "dusting" or spalling too rapidly. As a last resort in the case of severely spalled brick there may be no option but to replace the brick—a difficult, expensive (particularly if custom-made reproduction brick is used), and lengthy process.

Metals, other than cast or wrought iron, that have been pitted and dented by harsh, abrasive blasting usually cannot be smoothed out. Although fillers may be satisfactory for smoothing a painted surface, exposed metal that has been damaged usually must be replaced.

REFERENCES

Ann E. Grimmer, *Dangers of Abrasive Cleaning to Historic Buildings,* Technical Preservation Services Division, Heritage Conservation and Recreation Service.
Norman R. Weiss, *Exterior Cleaning of Historic Masonry Buildings* (Draft), Heritage Conservation and Recreation Service, Office of Archaeology and Historic Preservation.

VISUAL INSPECTION OF PAINTED SURFACES

The appearance of stains or peeling layers does not mean that the paint was the cause of the damage. The reason may be entrapped moisture, faulty preparation of the surface to receive the coating, paint that is incompatible with the surface, careless workmanship, runoffs from other material, mildew, or a variety of other causes.

Paint problems are found more commonly on exteriors than interiors. The exterior is exposed to temperature variations, rain, vapor pressure from within as well as blown abrasive debris striking the surface, and the destructive impact of solar rays. Interior paint surfaces must adjust to fairly minor temperature variations and with the exception of wet areas such as bathrooms and kitchens exist in a fairly benign environment.

Blistering. Bubbles in paint appear when water or solvent vapor is trapped underneath. Both types of blisters occur more frequently with oil and alkyd paints than with water-based coatings. They may be diagnosed by cutting the bubble open. If bare wood is exposed the blister was caused by moisture. If paint is found inside the blister was caused by solvent. This may have been the result of the paint surface drying too quickly, forming a skin which prevented the escape of the solvent and evaporation on the paint's surface.

Paint deterioration. (Photo by F. Wilson)

Wrinkling. This is usually evidence of paint applied too thickly. A film may form quickly and smoothly across the surface, but as the excess paint under the film dries, it decreases in volume. The surface film cannot shrink enough to accommodate it and sags into wrinkles.

Peeling. Coatings applied over dirt, grease or loose paint on wood or masonry that contain masonry may peel.

Alligatoring. A series of cracks that are usually caused by the inability of the top coat to bond smoothly to the paint below. Paint may have been applied incorrectly or may be incompatible with earlier coats. Serious alligatoring permits the seepage of water through to the wood or masonry.

Peeling from Masonry. Paint will peel from masonry or wood for identical reasons. Masonry, however, presents an additional hazard. Chemical compounds, alkalis, in masonry can work to destroy the paint adhesion to masonry units.

Chalk Stains. The paint may chalk as a result of the binder divorcing itself from the body of the paint; the binder is carried over the surface of wood or masonry by moisture.

Bleeding Knots. Knots may appear through paint due to the resin in them which is dissolved by the paint solvents and stains through to surface visibility.

Efflorescence. Efflorescence is a fine, fibrous collection of crystals projecting from the masonry surface. Paint may be destroyed by the appearance of efflorescence in masonry and concrete. Alkali compounds in the masonry are dissolved by moisture and carried to the surface. When the water evaporates, the compounds crystalize under the paint, pushing it away from the wall. This condition can be detected in older masonry near pipes, gutters, or downspouts as the result of moisture activating the compounds. If a wall has been in service for a number of years without efflorescence and then develops this problem, the cause is usually a recently developed condition that produced excessive wetting; the condition should be corrected before the wall is cleaned.

Green efflorescence occasionally forms on brick surfaces because of certain salts of vanadium contained in the clay used to make bricks.

Metal Staining. Shingles or other building elements may be discolored and stained by corroding metals. This condition is easy to detect. *Iron stains* are brown; they are caused by ferrous metal embedded in brickwork or other parts of the building so near to the stained surface that water has run from the metal onto the building surface. *Stains from copper and bronze* are bluish-green. They indicate a runoff of water from copper flashings and bronze fixtures.

Plant Growth. Moss, lichens, vines, and creepers are harmful to brickwork, whether it is painted or not. Particularly harmful are vines whose roots grow in mortar joints and eventually cause opening of the joints and retention of water on the masonry surface.

Mildew. Mildew and fungus will grow on paint. These molds will trap airborne dirt and some varieties eat through the paint. They are usually to be found in damp, shady areas.

REFERENCES

Canadian Building Digest, Division of Building Research, National Research Council Canada, ISSN 0008–3097.

Paint and Wallpaper, Time & Life Books, Alexandria, Virginia, 1980.

IX. A Procedure for Building Assessment

GENERAL METHODOLOGY*

Structural integrity is the key to building rehabilitation. Unless the basic structure is sound or can be economically repaired the full or partial demolition should be considered. Under unique circumstances when the structure is of historic importance the building may be eviscerated and an independent structural system erected within the peripheral shell, as was done with the White House in Washington, D.C.

Structural and mechanical systems evaluation is only one in the sequence of overall rehabilitation processes which includes social, economic, political, and technical considerations. Failure to recognize the importance of structural integrity in rehabilitation projects will lead to financial catastrophe, serious injury, or death to the building's occupants. In contrast, a reasonably well protected structural system and a well maintained mechanical system will safely serve a building and its occupants for considerable periods of time. The structural system may endure for centuries and the mechanical systems can be successively repaired and replaced over this period of time. It is reasonable and profitable, therefore, to take advantage of these potentials.

An outline of technical factors in building assessment and the sequence of their consideration follows.

1. Preliminary Evaluation Plan. Based on limited knowledge of the project to be undertaken estimates are made of staff, equipment, and expenses required to perform each of the nine basic tasks which follow.
2. On-Site Investigation. Reconnaissance is made of the building. This includes an estimate of the condition of the structure, sizes of principle members, and other salient features which will provide a groundwork sufficient to conduct a meticulous structural examination.
3. Off-Site Research. Documents relating to the building's life history, construction, and alteration are collected, copied, compared, and organized into a project dossier.
4. Preliminary Analysis. The materials gathered in on-site investigation and research are reviewed and analyzed to form the basis of a comprehensive evaluation. If major deficiencies which could warrant abandonment of the project are detected, they are brought to the attention of the project sponsor at this time.

---

* This material is adapted from a report prepared by Mr. Neal Fitzsimons, C.E., for the National Bureau of Standards entitled, "Structural Evaluation Guide for Building Rehabilitation."

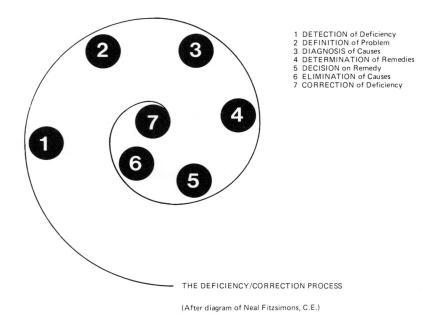

1 DETECTION of Deficiency
2 DEFINITION of Problem
3 DIAGNOSIS of Causes
4 DETERMINATION of Remedies
5 DECISION on Remedy
6 ELIMINATION of Causes
7 CORRECTION of Deficiency

THE DEFICIENCY/CORRECTION PROCESS

(After diagram of Neal Fitzsimons, C.E.)

The deficiency/correction process. (After diagram by Neal Fitzsimons, C.E.)

5. Final Evaluation Plan. The next five steps are planned and coordinated.
6. Laboratory Tests. Specimens collected from the building during the first and later on-site examinations are subjected to appropriate laboratory testing.
7. Field Tests. If analytical techniques cannot be applied with confidence to the structure because of unascertainable conditions, limited field tests can be used, to gage the structural integrity of in-place materials.
8. Additional Investigation. In-depth continuation of on-site investigation.
9. Additional Research. In-depth continuation of research.
10. Structural Evaluation. A comprehensive report is prepared summarizing conclusions concerning structural strength, stiffness, and stability, and giving detailed recommendations for the correction of any deficiencies is submitted.

Example—Structural Area

The evaluation of the strength, stiffness, and stability of the building requires an accurate description of its existing condition. This should be noted and carefully recorded in all its detail.

Such deficiencies as lack of rectilinearity of the structure or its aperetures is important. Rotation or translation of the structure, cracking, spalling, checking, or bowing of external walls should be noted. Broken windows or other openings which have allowed precipitation to enter the structure should be recorded. Such unprotected openings indicate that the interior environment has been subjected to extreme temperature and humidity variation in addition to the deleterious effect on both structural and nonstructural elements occasioned by water. Many external deficiencies can be detected by systematically scanning the building through binoculars. If more precise observations are required a tripod-mounted telescope or a theodolite can be employed.

Photography is a useful device to record the condition of structures. A telephoto lens helps to record inaccessible details and a wide-angle lens may be employed if visual obstructions interfere with views perpendicular to walls.

Deficiencies can be divided into four categories:

1. Distress. Manifestation of loading conditions that have exceeded the structure's inherent capacity, such as an office space used for heavy storage.
2. Deterioration. Caused by the inability of the structure to resist aggressive environmental agents such as termites in wood beams.
3. Damage. The result of structure subjected to extraordinary loads, as in the case of a vehicle crashing into a building.
4. Defect. A variation from the intended structural form large enough to significantly reduce structural capacity, such as a brick wall built out of plumb.

It is sometimes difficult to categorize deficiencies into a single type. For example it might result from an interaction of two or more types of deficiencies. The objective is to recognize the cause so that it can be corrected.

SITE INVESTIGATION TECHNIQUES

It is most important to insure that deficiencies are recorded in such a way that if patterns exist they will be revealed as early in the investigation as possible. It is therefore desirable to plot deficiencies systematically so that a continuous analysis for patterns can take place. In relatively simple cases, plans and elevations might be used as the basis of plotting patterns. If complications involving three-dimensional geometries occur, then the investigator might construct a Bristol-board or clear plastic model to plot deficiencies in a search for patterns.

With few exceptions standard surveying techniques can be applied to the investigation of most foundation problems. Differential levels and distances measured to the nearest one-hundredth of a foot coupled with angular measurements of one minute are normally sufficiently precise. Sometimes, however, it is necessary to use surveying techniques under special circumstances such as shaft plumbing or on tall buildings or when towers are involved.

Special equipment is required to measure potentially harmful vibrations. These might result from ordinary street traffic, from nearby railroads, or from reciprocating or rotating equipment located in nearby structures or in the building itself. Perhaps the most significant indicator of potential structural deficiency is the crack. Cracks can be measured as to width, length, and, when a more critical evaluation is necessary, depth. There are special techniques using dye penetration to reveal micro-cracking patterns in a limited area. However, a simple monocular reticule and feeler gages are usually quite adequate. Feeler gages, however, can be misleading when testing cracks in concrete members. Here cracking is often not vertically aligned with the surface but has a jagged profile. An alternative solution is to inject epoxy into a test area and then extract concrete cores and examine the crack patterns typical of a given situation. Without epoxy the cores tend to spall during extraction, destroying the crack pattern.

In the course of investigating deterioration in concrete an evaluation of the severity of the problem must be made. The investigator should select one or two small areas for evaluation and record them systematically with photographs and sketches during the removal of the deteriorated portion of the work until sound concrete is reached.

If the cause of damage is water it should be recalled that water can travel long distances and penetrate minor crevices and openings. It is sometimes necessary to employ special nuclear instruments to discover the water source.

Normally biotic problems are associated with timber. Dry rot is revealed using an awl. If termites are suspected a specialist should be consulted. A new technique using stress wave velocities appears to be practical for ascertaining the stress grade of timber members. However, regardless of mechanical or electronic devices used the investigator should systematically record the location and size of knots, splits, and checks. This is especially important in early timber construction, in which carpenters fit the framing members by mortising, tenoning, and notching. Splitting, a manifestation of shear failure, is detected in the undersurface of the notch. The stiffness of wood flooring can readily be determined by simply jumping up and down at the center of the span. Sagging is detected by placing a simple vial level at the quarter point of the span.

In general the key points for investigation are found in the area of high shear and low moment. It is in these areas that the greatest possibility of catastrophic collapse can be detected. In field erection of the structure the joints, bearings, and connections are normally the areas most vulnerable to defects. The investigator should therefore concentrate site investigation in these areas.

TESTING ALTERNATIVES

Field and laboratory testing is only required if there is no other reasonable alternative to obtaining accurate data or when it is prudent to confirm assumed or calculated values.

It might, for example, be less expensive to run a full-scale load test in a single bay that exhibits signs of deterioration than to define the extent of deterioration by examining each individual structural member. It is also useful to remove a small specimen of wood from timber members to determine species. This narrows the range of uncertainty concerning strength characteristics.

In general, full-scale tests are expensive and normally not conducive to repetition. They should be resorted to only if alternative means are impractical. Alternatives might include analytical interpretation and statistical interpolation of past tests, or less confidently extrapolation. The use of less than full-scale models in the same or analogous materials can also be considered in tests of structural components connections, and other elements. The results are then applied analytically to the entire structure. Similar structures in service can also be examined for analogous performance.

Regardless of the types of test selected, a practical analytical model should be developed in advance and be used in the planning and execution of all tests.

There are two types of full-scale structural load tests: nondestructive and destructive. Nondestructive tests may be divided into two classes: short-term loading tests and long-term loading tests. Short-term tests are normally used in rehabilitation projects; they may be either static or dynamic; dynamic tests may be either impact or cyclic.

The scope of full-scale structural tests normally includes a range for the magnitudes of loads, the number of loading cycles, the schedule of time and duration of load applications and removal, the load application frequency, the deformation range, and the environmental conditions under which the load test is carried out.

Structural testing using dynamic techniques has received increasing attention over the past decade, especially since transistors, digitizers, and other electronics have become available. However, it is unlikely that this technique will be practical for rehabilitation projects in the near future.

With the exception of long-term tests involving implanted instruments, the scope of full-scale structural tests is usually limited. It is often practical to only load one bay of a building, and this for one or two loading cycles. In nondestructive testing loads are kept well within the elastic limits. Because of these practical restrictions major preliminary analytical work should be carried out so that key data can be established within the scope of the test. These data permit a correlation between the selected analytical model and test results.

Scale-model tests are generally impractical for most rehabilitation studies. A highly unusual frame geometry might require such tests if the structure has relatively fixed joints. If this is not the case it is almost impossible, or at best very expensive, to model. In any event an analytical model should be established before any structural tests are undertaken.

Cost factors encourages the investigator to make a major effort to secure documentary evidence of a building's construction characteristics, rather than run tests. But this should not discourage verification tests when prudence dictates their necessity. The psychological trap in verification, however—giving only perfunctory attention to test results because they "look okay"—should be avoided.

It should be kept in mind that even a simple one-day deflection test without strain measurements could cost at least $1,500 and require two weeks for planning, execution, and reporting. The gathering of strain measurements involves more specialized staff and equipment. A bill of $5,000 to $10,000 can be incurred very quickly. If dynamic testing is used these costs might triple. (All figures in this paragraph reflect 1979 prices.) Concrete core collection and compression testing cost about $100 per core, but if petrographic examination is required another $100 to 500 can be anticipated. Steel coupons have similar unit prices for extraction, tension testing, and metallographic examination. Special physical/chemical tests would greatly increase these prices. Wood identification can be done for a small fee at the Forest Products Laboratory, Madison, Wisconsin, 53705. But strength testing of wood specimens has been a long-standing problem.

There are many testing techniques which determine the in-place strength of materials. Most of them rely on "hardness," a somewhat nebulous concept, and therefore use scleroscopic techniques which require the accurate measurement of rebound or indentations. Crude variations of this are found in tapping concrete, bricks, or steel with a hammer and judging quality by its "ring," pressing a thumbnail, awl, or knifeblade into wood to get a "feel for it;" or kicking the heel into the ground to check for "firmness." Frank Lloyd Wright is reported to have used the point of his umbrella for this purpose.

Where the superficial characteristics of a material are indicative of its structural characteristics these tests have some merit. This is true in certain well defined cases, as when gross deterioration has taken place on the surface or when core deterioration can be detected by plunging an awl through a veneer of sound material. Their usefulness is, however, limited. The Schmidt hammer, for example, has been in use for many years as a quick means for determining in-place strength of concrete. Experience and evaluation has clearly shown that it has limited value for this unless only a gross estimate is satisfactory. It can, however, give the investigator a "surface hardness" map of of a concrete structure from which potential weak points can be isolated and studied. Even in this case, only members of similar dimensions should be included in a given test series. Comparing measurements on the web of a relatively thin concrete joist with that of a large column would be totally inappropriate. Other tests such as the Windsor probe have similar limitations. On the other hand, ultrasonic testing, although much more expensive, is reasonably well accepted as a reliable tool for determining in-place characteristics of concrete and steel. A single day of ultrasonic testing by a trained team would probably cost about $500 (1979), excluding traveling expenses. If well planned, a 1–3 day program would be satisfactory for most rehabilitation projects.

A pachometer, which indicates the location of reinforcing bars in concrete, costs about $150–$200 per day (1979), including technician. Although a useful tool for simple rebar configurations, its reliability markedly decreases with more complicated ones.

STRUCTURAL EVALUATION

The final structural evaluation report should include a positive statement on the building's overall structural strength, stiffness, and stability in its as-is condition and, if remedial work is necessary, a similar statement on its post-remedial condition. In the latter case, the specific assumptions behind the results of the remedial effort must be defined.

Any remedial work should follow the deficiency/correction process outlined in the accompanying figure and the final evaluation report should cover each of the seven steps for every type of deficiency detected. It should be emphasized that step 6, elimination of the causes, is extremely important. Sometimes when this is impractical, the remedial work should be directed to minimizing vulnerability to a cause.

The structural evaluation report should interface with the fire evaluation report, both from the standpoint of collapse vulnerability and changes in apertures which may be required for emergency purposes. Further, new fire cladding and doors add dead weight to the structure, and sprinkler piping may require the drilling of holes in structural members.

This latter point must also be considered with respect to electrical, sanitary, mechanical, and communication conduits. Mechanical and electrical equipment can add large, usually eccentric loads. Further, rotating or reciprocating equipment can add dangerous dynamic or vibrational loads. The former will affect the stability of the structures while the latter can cause more subtle long-term reductions in structural strength at joints, particularly in old masonry structures.

New environmental control systems can also create problems by changing the ambient internal moisture/temperature conditions. It takes time for buildings to adapt to new conditions and in the process minor structural accommodations are made which can result in vexing cracks.

Foundations should not be overlooked in the structural evaluation. Added loads and changing drainage patterns or water table levels can impact on building foundations in detrimental ways. Over a period of time heating a hitherto unheated basement may cause moisture migration in the surrounding soil and change its mechanical properties. Paving over a grassy area changes drainage characteristics which, in turn, can affect the underlying soil. In general, overall settlements are far less onerous than differential settlements. There is also the classical case of lowered ground water permitting a change in the biotic environment of wooden piles, inducing rot and subsequent foundation failure. All three major modes of foundation failure, subsidence, rotation, and translation should be considered in the final report.

In the final evaluation report, and indeed, throughout the whole evaluation process, the investigator should keep in mind that structurally a building is a kinetic entity—constantly moving to adjust to variations in loads, weather, vibrations, etc. Its members and joints are restless in all modes of strain: expansion, contraction, dilation, and detrusion. This in turn causes other distortions such as translation, rotation, and differential motion among structural elements, but also between structural and nonstructural elements as well as among nonstructural elements.

As long as all these strains and movements are within certain limits and are not significantly progressive, the structure will satisfactorily function indefinitely, but systematic inspection should be carried out to insure these limits are safe.

In general, consideration should be given to diurnal and annual cycles as well as sporadic exposure situations.

Diurnal events such as the sun rising and setting influence crack patterns. The sunny side of the building expands much more than the shady side, causing a particular movement and strain pattern. Drying and dampening is also usually a day/night phenomenon superimposed upon a seasonal phenomena.

Annual cycles, such as heavy rains and snows, usually are associated with particular wind directions, so that one portion of a building suffers a more severe exposure than another.

Earth tremors (even discounting earthquakes) tend to be sporadic, but nonetheless can influence strain/movement patterns, as do vehicular traffic, sonic boom, etc.

Geometry must be considered, including both the gross geometry of the structure (manifested, for example, by racking) and the member geometry (such as section characteristics). In considering loads, magnitude, direction, and frequency must be determined. Aggressive or catalytic environmental agents should be watched for. Materials factors, resulting from the use of the wrong or defective materials, should not be overlooked. These and other factors are listed in the Table of technical notes.

TECHNICAL NOTES

1. *Loads and Agents*
   (a) Dead loads: Uniform, nonuniform, concentrated
   (b) Service loads: Static, dynamic, vibrational
   (c) Geophysical loads: Wind, snow, rain (ponding), earthquake
   (d) Geotechnically induced loads: Lateral pressure, Differential settlement, structural translation/rotation
   (e) Environmentally induced loads: Temperature, humidity, biotic forces
   (f) Environmental agents: Freeze/thaw, aggressive air, aggressive water, heat, moisture, biotic agents
2. *Qualities and Properties*
   (a) Construction materials
   (b) Geotechnical materials
   (c) Structural members dimensions
   (d) Structural member distortions, on and off site
   (e) General structure geometry
   (f) Construction workmanship
   (g) Connections
3. *Documentation*
   (a) Technical citations
   (b) Case dossiers
   (c) Engineering drawings and specifications
   (d) Construction drawings and modifications
   (e) As-built drawings (mech./elec. modifications)
   (f) Construction log books, photographs, sketches
   (g) Building operational log books
   (h) Building services reports
   (i) Media reports
   (j) Codes and regulations
   (k) Climatological reports
4. *On-Site Investigation*
   (a) General site examination
   (b) Condition survey
   (c) Sketches and photographs
   (d) Specimen selection and collection
   (e) On-site document collection
   (f) Owner/occupant interviews
   (g) Field testing
5. *Off-Site Research*
   (a) Laboratory testing: Specimens, joints, members, models
   (b) Documentation
   (c) Manual analysis
   (d) Computer analysis

6. *Structural Evaluation*
   (a) Primary physical deficiencies: Replacement of members?
   (b) Secondary physical deficiencies: Major repairs of members?
   (c) Tertiary physical deficiencies: Minor repairs of members?
   (d) Violations of codes and regulations
   (e) Structural capacity
7. *Rehabilitation Plan*
   (a) Foundation
   (b) Deficiency-free sections
   (c) Irreparable (Salvageable/nonsalvageable) sections
   (e) Rehabilitation recommendations
   (f) Structural capacity

As the project progresses better estimates of time and expenses can be made as actual data are accumulated. Also, schedules can be adjusted to suit staff exigencies or vice versa. The net result of using this project management system is that there is a well defined body of data collected on time and expense for each of the typical tasks associated with a rehabilitation project. These data are valuable for planning subsequent projects. Further, during the course of the project each staff member involved knows exactly who is expected to do what by when.

It will be noted that the process shown is a closed loop, in that data gained from any given project is fed back into the "system" so that future projects can benefit from present experience. This, at present, is an ideal situation which may be worthwhile considering.

Preliminary Steps

The most important beginning step for any structural evaluation project is the preliminary plan. Associated with this is the establishment of a project dossier, made up of a project notebook and a file which contains all correspondence pertaining to the project as well as collected reference material. The items in the project notebook are explained in detail below:

1. Journal Entry List. Used to record all important agreements, decisions, or actions made by the project staff. If the remarks concerning the entry are extensive, then an attachment should be made, noted under the remarks, and filed in the dossier with the journal entry number clearly marked.
2. Contact List. Designed for the orderly recording of people that the staff have contacted of should contact in the course of executing the project. For example, present or past occupants of the building, the owner or manager of the building, the architect/engineer, etc.
3. Document List. Used mainly to compile notes on books, periodicals, letters, plans, specifications, etc. which might aid in the structural evaluation of the building. It should not be confused with a citation form, which is normally filled out after a document has been procured.
4. Photo list. A record of all photographs taken or collected by project staff of the building and adjacent areas. It is absolutely imperative that negatives and positives both be carefully marked and recorded on this list.

5. Sketch List. Related to the photo list; the same general remarks apply. Among the first sketches made should be one which shows the camera positions of photographs appearing on the photo list. The sketches completed on the reverse sides of the general building information and member/assembly information forms should also be entered here, as should any other sketches made.

6. Specimen List. Required if any samples are to be taken in the course of the structural investigation. The sketches of the sample locations along with sketches and photographs of the samples themselves should be entered on forms 4 and 5.

7. General Building Information and Member/Assembly Information.

Once the project dossier is assembled in its initial form, the next major step is to collect the items of the investigation kit for the type of structure being studied.

At the same time the investigation kit is being assembled, information about the building and its site should be collected. Plot plans are usually available if a building is situated in a metropolitan area. They can be obtained from the county recorder's office, or the city clerk's office. Also, Sanborn maps can be obtained in most metropolitan areas. Sometimes older versions of Sanborn maps can be useful in ascertaining the past built-up environment in which the structure was situated. Local history sections of libraries and historical societies can be extremely valuable sources for maps, plans, photos, and documents on older structures.

Contents of Project Dossier

A. Project Notebook
  1. Journal entry list
  2. Contact list
  3. Document list
  4. Photo list
  5. Sketch list
  6. Specimen list
  7. General building information form
  8. Member/assembly information form
B. Project Files
  1. General correspondence
  2. Journal entry supplements
  3. Collected documents and citations
  4. Photographs and negatives
  5. Field reports and sketches
  6. Test results (field and lab)
  7. Analytical studies
  8. Evaluation report

Structural Investigation Kit

Camera, 110 (pocket type) with flash and film
Camera, 35mm (*f*3,5.59mm, or larger aperture) with flash and film
Lens, telephoto
Lens, wide angle
Tripod, camera
Recorder, cassette (pocket type)
Binoculars, 7 × 35mm or better
Monocular, 8 × 20mm
Magnifier, pocket style, 4 × to 10 ×
Mirror, inspection
Optical distance-measurer
Fiberscope, flexible
Reticle, millimeter, tenths scale
Level, hand
Level, circular
Level, vial
Level, carpenter's 4 ft
Caliper, inside/outside 5 in.
Rule, 6 in. with millimeter scale
Rule, folding 6 ft
Tape, 50 ft
Scale, engineer's
Scale, architect's
Gage, feeler
Flashlight
Penlight
Flexilight
Calculator, pocket
Hammer, masonry
Hammer, ball peen
Awl
Plumb bob
Screwdriver
Pen knife
Handbook, engineering (pocket type)
Project notebook with blank forms
Clipboard with blank, ruled, and graph paper
Pocket notebook (3¾ × 6¾ in.)
Bags, specimen
Tape and gummed labels
Felt markers and crayons
Heavy string
Tape, plastic
Tape, cellophane
Cards, marking (5 × 8)
Hairspray, aerosol
Penetrant dyes
Glass cover slips
Glue, cyanoacrylate or comparable
Hat, safety; Shoes, safety; Gloves, canvas
Increment borer

SITE INVESTIGATION PROCEDURES

Local building departments may be able to provide engineering drawings and site maps if the building is not too old. A recent survey of local governmental agencies in the greater Washington area indicated that plans would not be available at most building departments if the building is more than about 5 years old. Some departments do have microfiche format. However, the quality of the reproductions may be wanting. In any case, no building department contacted in the greater Washington metropolitan area required building plans to be deposited before 1952.

If a building has historical significance, the plans might be available through Historic American Building Survey or the Historic American Engineering Record, both of which may be contacted at U.S. Department of Interior, Heritage Conservation & Recreation Service, Pension Building at Judiciary Square, Washington D.C., 20240.

Of course, owners, building managers, or architect/engineers responsible for the original plans or modifications might well have plans for the building and should be contacted.

It should always be kept in mind, however, that it is rare to find any plans or specifications that accurately reflect the actual construction of a building. Therefore, a detailed on-site investigation of the building is invariably required.

In addition to the overall plan for a structural evaluation a more detailed sequence should be developed for the on-site investigation of the structure. This sequence will be highly dependent on the size and type of structure. The following general guidance is appropriate.

First, proper clearances and notifications must be given in advance so that the investigator will have ready access to the building upon arriving at the site. Normally the present owner, the building manager, or occupant should be included in this notification. Keys should be acquired prior to the site visit if possible. If a building is vacant, it might be appropriate to notify the local police and nearby residents.

Second, the evaluation of the strength, stiffness and stability of the building requires an accurate description of its "as is" condition. The site investigation procedures must result in enough information for the analysis of the structural capacity of the building as is, and must identify the causes of any deficiencies so that appropriate remedial work can be properly planned and when completed will be relatively permanent, that is, prevent the development of any future structural deficiencies.

The overall condition of the building should be noted on the general building information form. Such manifestations of the structural deficiencies such as lack of rectilinearity of the structure itself or its apertures is important. Similarly, any rotation or translation of the structure should be recorded. Cracking, spalling, checking, or bowing of the external walls should be noted, as should broken windows or other openings which may have allowed precipitation to enter the structure. Similarly, the roof should be checked for this possibility. The reason for this is that any water entering the structure can have deleterious effect on

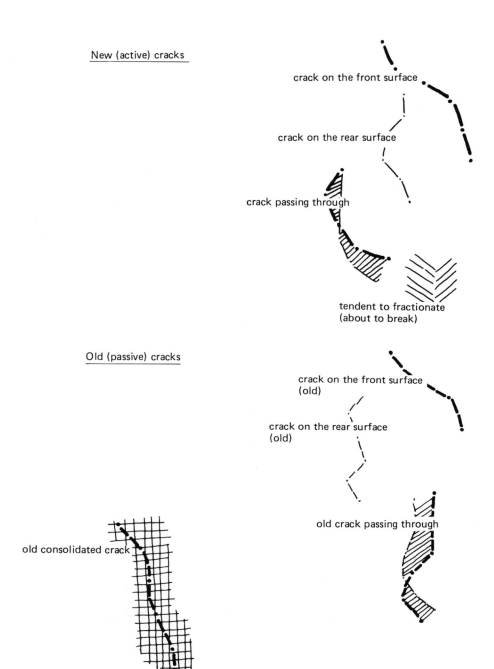

New (active) cracks

crack on the front surface

crack on the rear surface

crack passing through

tendent to fractionate
(about to break)

Old (passive) cracks

crack on the front surface
(old)

crack on the rear surface
(old)

old crack passing through

old consolidated crack

Symbols for stuctural cracking. (From "Research Support for Building Rehabilitation Studies in the Area of Strength" by Neal Fitzsimons, C.E.)

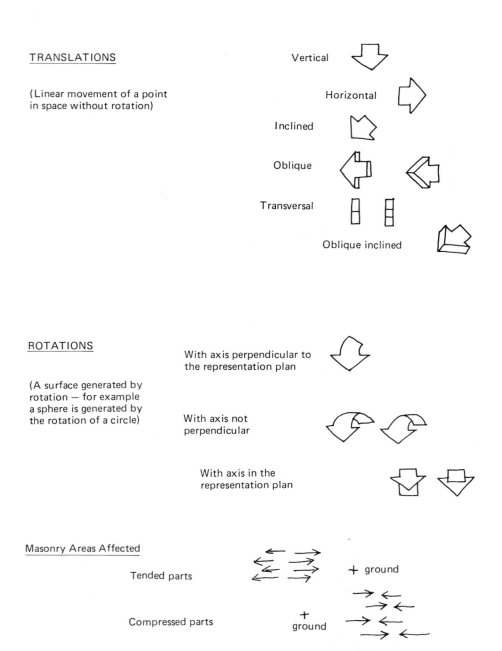

TRANSLATIONS

(Linear movement of a point
in space without rotation)

Vertical

Horizontal

Inclined

Oblique

Transversal

Oblique inclined

ROTATIONS

(A surface generated by
rotation — for example
a sphere is generated by
the rotation of a circle)

With axis perpendicular to
the representation plan

With axis not
perpendicular

With axis in the
representation plan

Masonry Areas Affected

Tended parts

+ ground

Compressed parts

+
ground

Symbols for structural movement. (From "Research Support for Building Rehabilitation Studies in the Area of Strength" by Neal Fitzsimons, C.E.)

both nonstructural and structural elements. Also, openings indicate that the interior environment has been subjected to temperature and humidity variation and concomitant adverse effects.

Many of these external deficiencies can be detected by systematically scanning the building through binoculars. If more precise observations are necessary, a tripod-mounted telescope or a theodolite should be used. Briefly prepared sketch elevations are necessary for (1) indicating the location and nature of the deficiencies, and (2) for referencing close-up photos and sketches of the deficiencies.

The use of a standardized set of symbols for noting manifestations of deficiencies and repairs is highly desirable, but unfortunately there is no available standard. However, a set has recently been proposed and appears to be a good basis for a universal system. (See accompanying figures.)

Photographs should be planned for the best time of day with reference to sun and shadow. Rooftop shots from adjacent structures can be very useful. Telephoto lenses are helpful in recording these shots, as well as details that are inaccessible. A wide-angle lens may be necessary if visual obstructions interfere with shots that are best taken perpendicular to walls.

Whenever possible, deficiencies should be categorized as one of four types: distress, deterioration, damage, or defect. Distress is a manifestation of loading

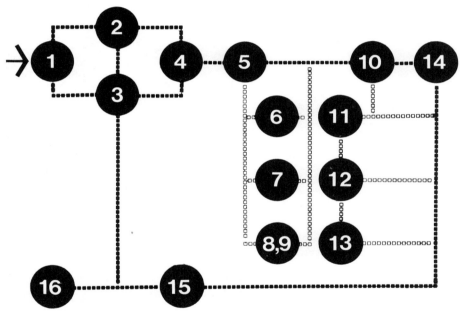

Optional Link ▫▫▫▫▫▫▫▫▫▫▫▫

Rehabilitation evaluation plan (structural evaluation & code)

REHABILITATION PROJECT MODEL
(Structural Evaluation)

1—Preliminary Evaluation Plan	10—Structural Evaluation
2—On-Site Investigation	11—Nonstructural Evaluations
3—Off-Site Research	12—Rehabilitation Plan
4—Preliminary Analysis	13—Rehabilitation Work
5—Final Evaluation Plan	14—Final Report
6—Laboratory Tests	15—Rehabilitation Data Bank—Citations/Cases
7—Field Test	16—Future Rehabilitation Projects
8–9—Additional Investigation/Research	(After Neal Fitzsimons)

conditions that have exceeded the structure's inherent capacity, e.g., office space used for heavy storage. Deterioration is caused by the inability of the structure to resist aggressive environmental agents; termites in wood beams would be a common example. Damage is the result of the structure being subjected to extraordinary loads, as in the case of a vehicle crashing into a building. Finally, a defect is a variation from the desired/intended structural plan which is large enough to significantly reduce structural capacity, such as brick wall being built out of plumb.

Unfortunately, it is often difficult to categorize a given deficiency such as a crack or spall into one of these four types. Sometimes, for example, the deficiency might be the result of the interaction of two or more types. The key idea is that the cause of the deficiency has to be recognized so that it can be mitigated or removed if extant or recurring.

X. Some Consideration of Joints in the Building System

Function of Joints

Principles of Wall Design

Performance Requirements

Construction Requirements

Construction joints, control joints, and movement joints are joints required for building the structure. The remainder of the joints in buildings might be described as architectural joints placed for the convenience of the architect or builder in limiting the size of building components for visual or functional reasons. All joints are separated joints, including the interlocking types, but that term usually refers to joints that occur as a result of deliberate separation of adjacent components.

FUNCTION OF JOINTS

Joint requirements will vary depending on the building type, the position of joints, and the expectations of the designers and users of the building. The joint function in general will probably include some measure of connection and environmental control. Connection will be required to ensure structural stability either of a single component or of a large proportion of a building. Panels may need to be connected together for structural continuity across joints, window panels must connect to wall panels, and both may connect to a third component.

Joints may be required to impede or control to some degree the passage of radiation, air, sound, heat and cold, water, vermin, and intruders. They are expected to maintain this control for a reasonable length of time. For convenience of construction, joints should be quick and easy to make. For the owner's convenience the joints should also be easy to maintain, if maintenance is required. They should, of course, be aesthetically acceptable.

PRINCIPLES OF WALL DESIGN

Most contemporary buildings are wall clad using panels of some sort or other. Even when traditional materials such as brick are used for modern buildings they are usually used as panels or a curtain or cladding material. This is important to remember, for building joints cannot be considered in isolation but must be treated as part of an overall design of the wall and other components.

The main elements of the wall, as advocated by the Division of Building Research of the National Research Council of Canada, are the following:

1. A structural air barrier as close to the interior as possible to control air leakage
2. Insulation where required for heat control, placed outside the air barrier and structural wall
3. Exterior cladding to perform aesthetic, solar, and rain-shielding functions as required, open at the back to a drained and pressure-equalized space.

Where it is possible to build such walls they would seem to have considerable advantage over other types of wall design, particularly for a demanding client. Since the seal for the wall is at the interior, on the warm side of the insulation, movements are likely to be small. It will therefore usually be possible to achieve a reasonable measure of tightness. Since no weathering function is involved, and the sealant used can be protected, it will probably remain intact.

Where this method of wall design is followed there is usually no need for achieving a seal at exterior joints, and in fact some joints must be deliberately left open to achieve pressure equalization, and drainage from the space behind the cladding. The requirements of such joints are not very critical; in many instances the joints contain no filler materials. Overlapped units can be used and tolerance correspondingly eased.

When air space is not provided in the exterior wall design, the principles delineated can still be applied. The air seal must still be applied to the structural wall inward from the insulation. Exterior wall panels can be placed directly against the insulation. If they are precast, some space may still exist between the panels and the insulation, and it may be sufficient to provide some measure of pressure equalization. The vertical joints will normally need to be closed to prevent direct entry of rain, but horizontal joints can be left open if they are designed to drain.

When sandwich-type panels are used they provide all the barrier requirements within the panel. The requirements of the joints then become much more critical, as they are now also required to provide all the barrier requirements. It is still possible to apply the principles relating to air leakage and rain penetration at the joints by sealing at the inside of the panel, and providing rain deterrents and drainage at the outside. Sealants or gaskets are usually required, and the tolerances involved in the panels may make it difficult to achieve a satisfactory detail.

PERFORMANCE REQUIREMENTS

Joints must control heat flow, air flow, water vapor flow, rain penetration, light, solar and other radiation, noise, and fire.

The transmission of heat through the joint, either in or out, should be similar to the rate for adjoining panels. If this varies, pattern stains may develop, or in cold weather localized condensation may be caused by the cold bridge effect across the joint. In systems that utilize continuous insulation behind the panels or behind a pressurized and drained air space this is not a problem. For sandwich-type panels this can be a potent source of problems from thermal bridge effects at the joints, as already noted, and panel warping from thermal and moisture effects.

Air flow across a joint from inside or from outside could be a cause of joint problems. If the air moves from the outside in, it can cause rain penetration; if from the inside out, condensation in the joint. Both will be avoided if an adequate air seal is achieved. For rain screen systems, such a seal is usually achievable on the backup or structural wall. For sandwich-type panels, the seal must be made from panel to panel and this may not be easy to accomplish.

Water vapor will be carried into the joint by vapor or air leakage, and the danger of condensation will be related principally to the effectiveness of the seal. Whether condensation takes place will depend on the temperature of the air and the relative humidity. In the case of rain screen systems it is unlikely to be a problem since the moisture can usually travel harmlessly to the outside.

Rain penetration will depend on the amount of driving rain that impinges on the building surfaces and how this is forced to flow over the building wall face. If rain runoff on the building is concentrated at vertical joints due to the cladding design, it may be very difficult to avoid penetration of water and deterioration of sealants or gaskets in such joints. The building façade should be designed to direct water runoff away from critical joints, and the joints designed to deter water penetration into the joint, with sealing of the joint located as far inward as possible. As some penetration will undoubtedly take place at some time, allowance must be made for the water to get out.

All walls of most buildings receive a certain amount of ultraviolet light. Rubber and plastics used as sealants and gaskets can be severely deteriorated by the ultraviolet unless ultraviolet absorbers and pigments are incorporated in the materials. Radiation on the panel and variations in ambient temperature can change surface temperatures, which may result in considerable movement of the panels. This movement can cause damage to panels made from composite materials, at joints between panels, or at junctions of glass and panels, unless taken into account in design.

Information regarding the acoustical behavior of exterior wall joints is rather limited, but one can assume that there must be an airtight seal somewhere from inside to outside. Since this is also a requirement in relation to the other control items, if the other requirements are met, the sound control probably will be as well. It will probably be of greater consequence in relation to joints between partitions and exterior walls, or between panels of walls inside the building for noise transmission between rooms.

The fire resistance of joints has not been given much consideration, but it seems reasonable to suggest that joints should have the same fire resistance as the walls of which they are part. In this regard, however, glass does not provide any resistance to fire, so perhaps this characteristic should not be required of joints. In the event of a fire, the thermal expansion of building panels could damage or destroy the jointing material and cause the panels to pop off, resulting in the collapse of the wall, or the sealing material would melt or burn away to leave gaps through which the fire could spread. In some systems with an air space behind the cladding, there may be danger of the space acting as a "chimney" for spread of fire. This probably cannot be assessed separately, but must be considered in the context of the complete wall.

General performance requirements include resistance of jointing materials to the action of fungi, microorganisms, insects, birds, and rodents. They need to be resistant to damage by the public using the buildings, particularly at the lower floors. This is particularly necessary if the sealant materials used are exposed in the joints.

The design of seals should facilitate maintenance if it is required, and it most likely will be. Repairs and replacement of the jointing material, or of complete building panels should be relatively simple and cheap.

CONSTRUCTION REQUIREMENTS

Construction requirements for joints have to do mostly with efficiency and productivity at the building site. The designers' considerations should relate first to whether the following can be achieved: the required degree of air tightening on the inside or weather tightening on the outside. If the method of erection makes large demands upon the skill of the workmen, it may be that such demands exceed their capabilities, and it is unlikely that a sufficient degree of perfection will be achieved. The designer should also be concerned about the erection methods, so that construction proceeds in as simple a manner as possible.

The design of the edge profiles of cladding panels should be such that they can be transported without damage and without the need for expensive protective packing. The edge profiles should allow handling or stacking on site, and positioning on the building without deterioration from normal wear and abrasion. If excessive care is required, units will be damaged or the work efficiency of the tradesmen reduced.

The joint must be designed in such a way that it can be assembled rapidly, and not require much preparatory work on the site. It is desirable that cladding panels require little or no special equipment for erection, that their installation require little skill, that the assembly be dry, and that it can take place under any weather conditions. During positioning and attaching of cladding panels, movement should be possible in three planes: vertically in the plane of the façade, horizontally in the place of the façade, and horizontally at right angles to the façade. In addition, rotation within any of the three planes should be possible.

REFERENCE

Baker, M. C., "Recognition of Joints in the System" in *Cracks, Movements and Joints in Building*, National Research Council of Canada (1972), NRC 15477

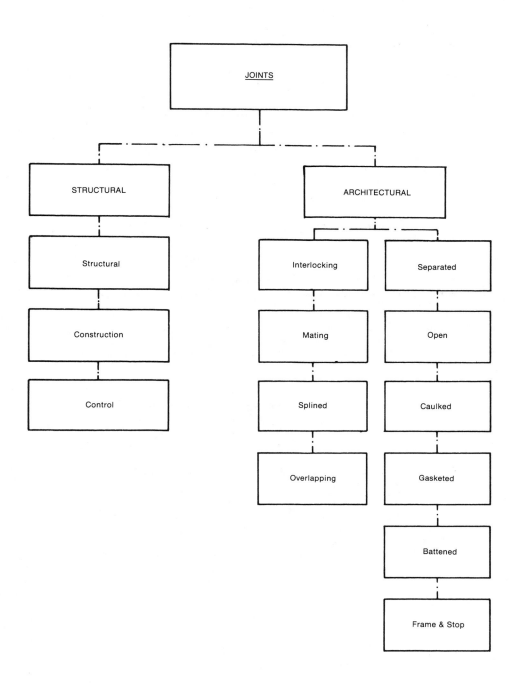

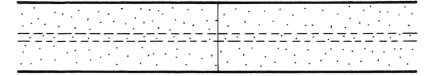

Construction joints or pour breaks in concrete are convenience joints that are designed to create a separation between consecutive concreting operations. The joint is the junction produced by placing fresh concrete against the surface of hardened concrete. No movement is expected and the reinforcment, if any, is continuous across the joint. It is often a vulnerable place for water leakage, particularly if concrete shrinkage or other building movement causes it to open.

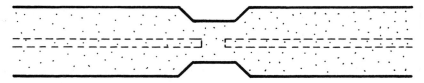

Control joints are not joints but the deliberate weakening of a section of construction so that cracking will occur along a line in a preselected location. Nature, however sometimes ignores these measures and cracks develop between control joint locations. If successful, these become separated joints.

SEALANT

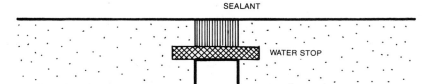

WATER STOP

Movement joints are complete breaks deliberately placed through floors, walls and roofs to allow for expansion and contraction. Contraction joints are intended to allow for shrinkage and to relieve tensile stress.
Expansion joints serve to eliminate or reduce compressive stresses that can develop as a result of thermal or moisture expansion.

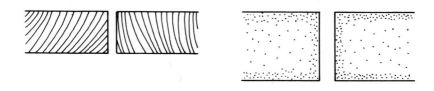

OPEN BUTT JOINTS

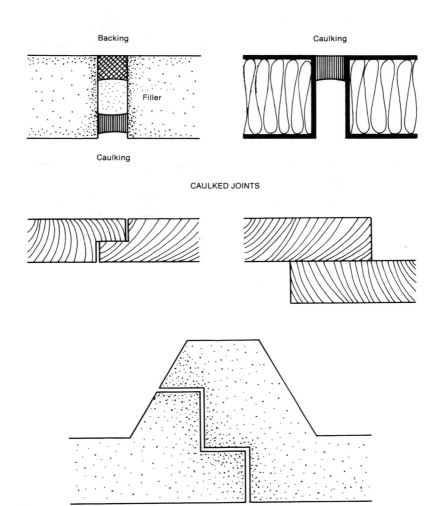

Backing

Caulking

Filler

Caulking

CAULKED JOINTS

OVERLAPPING JOINTS

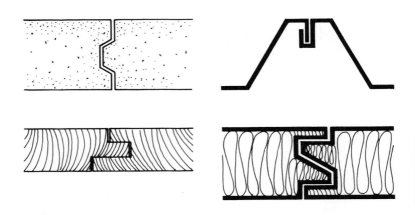

INTERLOCKING AND MATING JOINTS

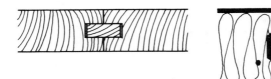

SPLINED JOINTS

XI. Test Methods for Existing Mechanical Systems

Plumbing Systems

    Physical Inspection—Plumbing System Components

    Water Tests—Water Supply Lines

    Water Test—Water Storage Tank

    Water Test—Drainage and Vent Systems

    Air Test—Drainage and Vent Systems

    Discharge Test—Drainage and Vent Systems

Electrical Systems

    Visual Inspection—Electric Branch Circuits

    Megohm Test (Megger Tester)—Electric Branch Circuits

    Voltage Determination—Electric Branch Circuits

    Circuit Analyzer—Electric Branch Circuits

    Circuit Breaker and Resistance Tester—Electric Circuit

Heating, Ventilating, and Air Conditioning Systems (HVAC)

    Pressurization—Building Envelope

    Tracer Gas Technique—Building Envelope

    Thermography—Building Envelope

    Humidity Indication by Color Change—Building Interior

    Electrical Impedance Hygrometer—Building Interior

    Electric Co-Heating—Building Interior

    Physical Inspection—Furnaces and Boilers/Heating System

    Ultrasonic Thickness Gaging—Containment Vessels

PLUMBING SYSTEMS

The implementation of a design complying with the code requirements in new construction is fairly straightforward, since inspection and testing of critical elements at key times is feasible as construction progresses and before the elements are concealed. However, this is not possible where evaluation of existing, largely concealed plumbing is required. Frequently, this difficulty has resulted in the tendency to totally replace aged plumbing rather than selectively rehabilitate existing systems. With modern techniques for evaluating plumbing systems, however, total replacement may possibly be avoided in favor of supplementing the existing system or replacement of deteriorated components only. Descriptions of some of the inspection and test methods available for evaluating the condition of existing plumbing systems follow.

Physical Inspection—Plumbing System Components

Safety Functions

A thorough inspection of the plumbing system should be carried out by a qualified plumbing inspector to gain a professional opinion on the condition of the system. The plumbing system is intended to function in a safe manner for the supply of potable water and the effective removal of waste products through sanitary drainage.

Physical inspection should include the following:

1. Conditions which could cause an unsafe building environment
   a. Excessively hot water (absence, inadequate design, or malfunction of temperature control devices)
   b. Explosion hazard due to absence, or defectiveness of, temperature and pressure-limiting devices
   c. Plumbing wall construction which will allow fire spread and passage of smoke or toxic gases in case of building fire
   d. Water or drainpipe leaks, or leaks around fixture connections which could allow gradual deterioration of structural elements of the building.
2. Conditions which could cause an unsanitary building environment
   a. Leaks in sanitary drains and vents (these leaks can contribute to the saturation, etc.)

b. Overflowing of sewage from fixtures onto floor or under/behind cabinets, etc.
c. Emission of sewer gas or suds from fixtures
d. Inoperative fixtures due to clogged drains or inability to obtain water
e. Inoperative water heater, or inadequate supply of hot water
f. Poor flushability of water closet
g. Absence of potable water supply, or supply of inadequate quantity
h. Backflow hazard (e.g., submerged fixture supply fitting outlets below flood rim, WC ballcocks without vacuum breakers, flushometers without vacuum breakers, dishwashers without air gaps or air breaks, etc.)
i. Absence or inoperability of essential plumbing fixtures and plumbing services.

Advantages and Limitations. Physical inspection provides quick feedback on the identification of problems, but the value of the inspection may be diminished in cases where major portions of the system are concealed. Many of the features identified in the inspection listing above are items that must be accessible, so inspection can be made with little trouble. However, accessibility will be difficult or impractical in some cases. Personal experience will be of value in forming an opinion on condition through physical inspection. The individual who has designed, repaired, and observed the dismantling of older systems should have an advantage in assessing the potential remaining useful life of the existing system or its ability to accept modifications.

Water Tests—Water Supply Lines

Leaks

A simple method to check for the existence of leaks in the water supply lines of the building:

1. Close all the valves on plumbing fixtures on the supply line to be checked.
2. Be sure the main water supply valve (or valves) is open.
3. Listen for a gurgling or murmuring sound in the supply pipe.
4. If no leak can be determined to exist in the interior supply line, and a leak still is suspected, listen for signs of leakage in the supply pipe leading from the water main outside the building and call the water company to check that line for leaks if gurgling sounds are heard.

Advantages and Limitations. The method is fast and simple. The results are immediate when leaks are large enough to be heard and when they occur in pipes which are not embedded or concealed.

Adequate Water Pressure

Clogged water supply pipes often cause a reduction in the water pressure at the faucets. One way to check the adequacy of the pressure is as follows:

1. Open the top floor sink faucets.
2. Open the bathtub faucets and flush the toilet. If the flow of water in the sink slows to a trickle, the piping may be of inadequate size or badly clogged with scale.

Advantages and Limitations. This is a fast and simple test method with results that can be obtained immediately.

Water Test—Water Storage Tank

Adequacy

If the water supply is from a well, turn on as many faucets as possible and observe the water after letting it run for ten minutes. If the water becomes muddy, the storage tank probably is undersized for the demand of the building. The capacity and recovery rate should be indicated on the nameplate mounted on the tank. An electric heater should have a capacity half again as great as a gas-fired heater to compensate for its slower recovery rate. For either type, check to be sure there is a pressure relief valve installed for safety.

Advantages and Limitations. This is an inexpensive, fast method for determining in a very general way the adequacy of the size of the water storage tanks.

Water Test—Drainage and Vent Systems

Leaks

The water test can be used to test for leaks in a drainage system. It can be applied to the entire system or to sections. If applied to the entire system, all openings in the piping should be tightly closed (except the highest opening) and the system filled with water to the point of overflow. The same conditions apply if the system is tested in sections, but no section should be tested with less than a 10-foot head of water. The water should be kept in the system, or in the portion under test, for at least thirty minutes before inspection starts; the system should then be tight at all points.

Advantages and Limitations. The water test for leak determination is relatively inexpensive. Leakage of water, should a leak occur, could cause some cosmetic damage in the existing building. Because the total drainage system seldom will be fully exposed for easy inspection, the point(s) of leakage may be difficult to determine accurately. The test method should indicate the general area of failure when leakage occurs. Some exploration through the removal of building components may be required to pinpoint the exact location and carry out repairs of the pipe or joint. If severe leakage occurs, or leakage is in an inaccessible location, consideration should be given to replacement of system components or possibly the total system.

Air Test—Drainage and Vent Systems

Leaks

The air test can be used to test for leaks in a drainage or vent system; it should only be applied to the entire system. An air compressor or other testing apparatus is attached to any suitable opening and all other inlets and outlets to the system are closed tightly. Air is forced into the system until there is a uniform gage pressure of five psi or 10 in Hg. This pressure is held without introduction of additional air for a period of at least 15 minutes. Any reduction in pressure indicates the presence of a leak in the pipe.

Advantages and Limitations. The air test for determining leakage is a rather simple and inexpensive test and requires little skill. A source of compressed air and adequate controls are required to ensure safe operating conditions and to guard against overpressurization of the system. The test procedure will give a quick indication of system quality with regard to leakage but the source of leakage may require additional testing (such as the soap and bubble test or listening for escaping air at the point of leakage). It also is possible to charge the system with gas so that leaks can be identified through the use of gas detectors. Because most drain and vent systems are enclosed in inaccessible construction, the source of leakage may be difficult to locate.

Discharge Test—Drainage and Vent Systems

Siphonage of Traps

To test for siphonage of fixture traps, fill one or more fixtures to an overflowing level and discharge them, then measure the trap seal retention. Trap seal retention should be at least 50% or 1 in. (25 mm), whichever is greater, in all traps when tested for self-siphonage or for cumulative effects of induced siphonage. The seals in the traps should be measured before and after each of several repeated discharges. If excessive siphonage occurs, the discharge pipes should be enlarged, cleaned, or otherwise modified. The number of fixtures to be discharged simultaneously in testing for stability of the trap seal is dependent upon the number and type of fixtures on the drain stack and their spatial arrangement.

Advantages and Limitations. The indicated procedures to determine trap seal retention have been used for years and are a fairly good indication of anticipated performance. Judgment must be used in looking at alterations of existing systems which may be required to carry unusual loading. Because of the health and safety issue of sewer gas emission with reduced trap seal, a conservative selection of simultaneous loading (extra heavy load) should be chosen when future use conditions are uncertain.

General Comment. Where future stories are proposed to be added to a building, the water piping systems should be tested to reflect the added static pressure anticipated to be imposed by the added height.

ELECTRICAL SYSTEMS

Since most electrical wiring is concealed inside walls, a complete inspection of all the parts of the electrical system in a building may not be feasible. Some general items, however, are easily checked which can aid in determining the electric shock and fire safety aspects of the system, as well as detecting functional and energy loss problems which may exist. Nearly all aspects of the National Electric Code in some way address the requirements for shock and fire safety. In older buildings there are potential problems through system overloads as new appliances and oversized fuses may be placed in the circuits.

The following are examples of evaluation techniques for assessing the condition of an electrical system in an existing building. While the evaluation methods mentioned in this section generally are recognized as being good practice, local codes and local officials should be consulted to determine whether any specific requirements apply to the individual geographical areas or type of building being evaluated.

Visual Inspection—Electric Branch Circuits

Circuit Faults

One of the best techniques available for assessing the condition of existing electrical systems is visual inspection of the wiring system. This technique helps to determine the condition of the insulating material and the methods used in making connections within the circuit. When analyzing existing systems, it is important to determine not only the condition of the system but also how the system is used during normal operation. During inspection it is important that the following items be noted (if they occur):

1. Overfusing. Check to see that the proper size fuse or circuit breaker is used in the branch circuit. Frequently, a fuse having a greater capacity than the original design capacity will be placed in the circuit and will create a potential overload condition in the circuit. This could result in ignition of material near the conductors.

2. Overlamping. Observe the wattage of lamps to determine if they are within the recommended limit. This is especially critical in confined spaces such as recessed lighting fixtures. With oversized bulbs in confined spaces the potential for overheating and resulting ignition of nearby combustibles is greatly increased.

3. Size and Number of Conductors. Knowing the anticipated load, it is easy to determine if the proper size wire and fuse combination is being used. The National Electrical Code should be consulted to determine if the existing circuit has the required capacity. Check for the existence of an adequate number of circuits with the proper voltage and current ratings for any anticipated retrofit equipment. Some common symptoms of circuit overloading include: (1) presence of many 20 or 30 ampere fuses; (2) the smell of burning insulation near the panel;

(3) fuses that have glass tops that are warm; (4) discolored copper contact points under the fuses; (5) overheating of wires as evidenced by melted insulation, and, in extreme cases, fire around outlet boxes, switches, etc., or wherever thermal insulation is tightly packed around wires; (6) fuses that blow frequently or circuit breakers that trip frequently; and (7) presence of surface-mounted lampcord extension wiring or multiple cords plugged into a single outlet.

4. Type of Conductor. Examining the conductor will indicate its type (copper, aluminium, or copper clad aluminum), and the wiring method (nonmetallic sheathed cable, armored cable, electric metallic tubing, rigid conduit, or flexible metal conduit).

5. Type and Condition of Insulation. Generally, the insulation around the conductor will be either rubber or plastic. The insulation should be physically examined to determine its brittleness. If there is any flaking or if any of the insulation falls off when bending occurs, it is deficient and should be replaced. An easy place to check for deficient insulation is in the exposed areas of the basement, around the circuit breaker or fuse box, and inside outlets or switches. Be sure to turn the power off to the circuit before removing outlet or switch plates or before probing any wires.

6. Connectors. Check for poorly functioning connectors which could cause excessive power losses and which might be a potential source of ignition. Power losses and low voltage can impair the functional operation of appliances and shorten the life of critical components.

Advantages and Limitations. Visual inspection provides immediate feedback on the condition of those parts of the electrical branch circuit that are accessible for inspection. To gain the most benefit from visual inspection it is best to have an experienced individual who has repaired or examined the dismantled older systems to gain first hand knowledge of potential problem areas. Visual inspection may detect only the most obvious defects because of the surface-only type of evaluation. It is recommended that only experienced personnel conduct electrical inspections.

Megohm Test (Megger Tester)—Electric Branch Circuits

Deteriorated Insulation

If the insulation appears to be deteriorated (crumbling or cracking), it is important to either replace the wiring or check the condition of the insulation with a megohm test performed by an electrician. In this test, the wires for each circuit are disconnected at the power supply end, and a Megger tester applies a test voltage to the wires at the other end. Branch circuits should read at least one megohm to ground. If lights or appliances are connected, readings should be at least 500,000 ohms. (Feeders should be tested in accordance with Article 110-20 of the National Electric Code). If there are any indications of the shorts in the circuit, the insulation is faulty, and the circuit wiring must be replaced.

Advantages and Limitations. This procedure for evaluating the condition of the wiring insulation is relatively easy, fast, and inexpensive. It is recommended that only a qualified electrician perform the test since high voltage is used and there are certain hazards associated with the test.

Voltage Determination—Electric Branch Circuits

Excessive Voltage Drops

The measure of voltage drop in a branch circuit is a good indicator of excessive impedence (e.g., excessive length) of the circuit. Conductors in branch circuits will provide a reasonable efficiency of operation if the following conditions are met:

1. The voltage drop at the farthest outlet of power, heating, and lighting loads does not exceed 3%.
2. The maximum total voltage drop on both feeders and branch circuits to the farthest outlet does not exceed 5%.

If voltage drops much in excess of these values occur, the circuit needs replacement. Generally, available voltmeters can be used for making the determination of voltage drops in the circuit.

Advantages and Limitations. The procedure for making voltage drop determinations is relatively easy and inexpensive. The conventional, commercially available voltmeter can be used. Measurements must be made under load with current also being measured. It is suggested that only a qualified individual, knowledgeable of electricity and branch circuitry, make the required measurements because of the danger of electrical shock when untrained personnel work with these potentially dangerous levels of electricity.

Circuit Analyzer—Electric Branch Circuits

Circuit Faults

A commercially available electrical device has been developed for analyzing potentially hazardous circuits. The analyzer is designed to determine various faults in existing wiring systems. It is intended for testing newly wired circuits, but also may be useful in assessing the proper performance of existing systems. The circuit analyzer is said to be capable of checking the following circuit conditions:

1. Open ground
2. Open hot
3. Open neutral
4. Hot/ground reversed
5. Hot/neutral reversed.

Depending on the type of analyzer, one or more test lights will be activated if the outlet is functioning and safely grounded. If ungrounded outlets are discovered, they should be grounded by installing a ground wire (this is especially important in kitchens and bathrooms). The manufacturer's instructions for the particular receptacle analyzer being used should be consulted for the detailed meaning of the lighter indicators. The analyzer can be used on old two-wire circuits with a special adapter.

Advantages and Limitations. Measurements obtained through the use of the circuit analyzer indicates only design or operating faults such as improper connections. This method of analysis will not help determine the condition of materials which make up the branch circuit. The technique is simple and quick, and the analyzer is commercially available.

Circuit Breaker and Resistance Tester—Electric Circuits

Condition of Circuit Breakers

A commercially available circuit breaker and resistance tester is claimed to provide the capability of evaluating circuit breakers by simulating an overload condition and proof testing the circuit breaker. A component of the test set allows evaluation of resistance of motor windings and other insulation resistance in the range from 1 megohm to 1000 meghohms. The current is limited to 5 milliamperes short-circuit current for maximum safety. The test set provides for quick assessment of the condition of electric circuit breakers to make sure that they operate in the intended range for overload protection. The resistance tester is intended to indicate potential problem areas relative to breakdown in electrical resistance of insulating materials.

Advantage and Limitations. Testing of the circuit breakers requires that the circuit breaker be removed from the panel. Because of potential danger in working with electric circuits, a qualified electrician should be engaged to conduct the tests. In older buildings which may not have a master disconnect, the utility company may be required to cut off service to the building during the removal of the circuit breaker.

HEATING, VENTILATING, AND AIR CONDITIONING (HVAC)

Several techniques have been developed which can help evaluate the performance of existing HVAC systems. Some of these are presented in this section. Also discussed are methods for evaluating the performance of the building envelope which contains the conditioned space. Techniques for measuring air leakage, as well as heat gains and losses, through the exterior of the building are valuable because space conditioning without proper humidity control can cause serious problems of condensation within the building structure.

A discussion of techniques for assessing the condition of the existing HVAC system and the envelope of the building follows.

Pressurization—Building Envelope

Air Leakage

Pressurization of the space within a building, either positive or negative with respect to atmospheric pressure, is one means of gaining a quantitative measure of the air flow rate. Pressure sensors, along with temperature measurements, are needed to make the determination of leakage rate for the given pressure differential. Generally, a window or door opening can be fitted with a plywood partition through which the air required to pressurize the structure is introduced. The air is allowed to seek points of penetration in an effort to equalize the pressure between the space and the surrounding atmosphere. The amount required to maintain the pressure differential provides an indication of the overall performance of the structure with regard to air leakage. It is possible to isolate contributing points of leakage by selectively taping cracks and joints during a sequence of tests. Internal air leakage paths can be identified using a fan to generate negative pressures and infrared scanners to trace the path through the construction.

Advantages and Limitations. The technique is rather sinple and not necessarily expensive (expense does vary with the complexity of measurements). The results using the air pressurization technique are not necessarily transferable to gain an indication of infiltration rates under natural conditions. The results provide a relative measure for assessing the structure to determine where within a range from good to bad the structure falls. The technique can be used effectively as a quality assurance procedure.

Tracer Gas Technique—Building Envelope

Air Leakage

The actual air exchange rate within a structure can be determined through the use of a tracer gas which is released into the space to be measured, then the concentration of the gas is monitored over a period of time. The exchange rate can be expressed through the following relationship:

$$I = -(1/t) \ln (c/c_0)$$

where: I = air volume changes per hour
  t = time
  c_0 = concentration of tracer gas at time zero (beginning of test)
  c = concentration of tracer gas at a specific time within the test period.

When the natural logarithm of the relative concentration ($\ln (c/c_0)$) is plotted as a function of time, the rate of infiltration (I) is the negative of the slope of the best fit line through the data points. Several tracer gases, such as carbon dioxide, helium, nitrous oxide, ethane, methane, and others, have been used as indicators of concentration decay. Currently one of the most popular is sulfur hexafluoride.

Samples of the air within the structures are taken and analyzed on the spot or stored in sample bags during the test and later analyzed.

Advantages and Limitations. The tracer gas technique is probably the best measure of natural infiltration in that it does not introduce any unnatural constraints for making the determinations, other than some form of forced air circulation to maintain mixing of tracer gas and air. The use of sulfur hexafluoride (SF_6) allows the use of very small amounts of harmless gas, generally on the order of 10–50 parts per billion, to be introduced into the space, thus altering to a negligible extent the composition of the air. At infiltration rates of greater than three air changes per hour, the tracer gas technique may give unreliable results because of the introduction of large amounts of unmixed air causing significant scatter in the data. At the higher rates of air exchange, the need for rapid sampling rates may also pose a problem.

Thermography—Building Envelope

Heat Leakage

Sources of high heat losses can be determined through the use of thermographic techniques. Thermography is a technique of portraying an object using the thermal energy radiating from the surface of the object. The energy radiated from the object (infrared region of the spectrum) is displayed on a cathode ray tube (either a black and white or color television monitor). On a black and white monitor, the intensity of the gray scale is proportional to the temperature of the object—within a given range, colder objects appear dark and hotter objects appear light. On a color monitor, the gradation of colors of the image indicates the surface temperature variation. Thus, since heat loss or gain leads to surface temperature variation, it is possible to detect areas of heat leakage through observations of the building's surface temperature. A photograph can be taken of the image on the screen of the monitor and a record of the thermal image can be obtained in the form of a permanent thermograph.

Humidity Indication by Color Change—Building Interior

Moisture

A rather simple, yet inexpensive, method of monitoring relative humidity within a structure is through the use of the color change humidity indicator. One such indicator uses cobaltous chloride as the basic ingredient for indicating the relative humidity. Blotting paper is used as the holder of the indicator. As the cobaltous chloride is exposed to the atmosphere and equilibrium is reached, its color changes. A range of relative humidity indication from 10 to 80% can be obtained. There are plug-type indicators that may be used to indicate the relative humidity in wall cavities or enclosed roof ceiling cavities. The color change indicator can also be used to monitor the relative humidity of the occupied spaces.

Advantages and Limitations. The color-change type indicators are inexpensive and are relatively easy to use. The accuracy of the indicator can be affected by temperature changes and long-term exposure to high humidity, high temperature, or direct sunlight. As an indicator of acceptable levels of relative humidity in a space within a building or an unacceptable condition within a structural compartment, the color change indicator should perform adequately. Direct contact with water will cause the chemicals to leach out of the paper and thus lose calibration. It is possible to obtain an indication of relative humidity within 5% of the published equilibrium point. A 10°F change in temperature will affect the indicator accuracy by 2.5%. A limitation is that the equilibrium point varies with the specific chemical used, and it is a specific point (not a range).

Electrical Impedence Hygrometer—Building Interior

Humidity

With a change in humidity, many substances absorb or give up moisture and exhibit changes in electrical impedance. Sensors can be produced having dual electrodes or windings which are electrically separated by a thin film of binder material containing a salt solution (such as lithium chloride). Means are provided for measuring the AC electrical impedance between the electrodes through the salt film. The impedance through the salt responds to changes in relative humidity and temperature, so that accurate indications of relative humidity can be obtained.

Advantages and Limitations. The electrical impedance hygrometer is relatively sensitive to humidity changes and provides the capability for remote read-out. Precision on the order of $\pm 1.5\%$ RH can be obtained and elements can be produced to cover the range from 10 to 90% RH. The electrical impedance hygrometer is susceptible to damage by air contaminants and water. Frequent calibration checks are required. The elements can be placed in remote places for monitoring changes in relative humidity in an effort to gain information on excessively high moisture conditions.

Electric Co-Heating—Building Interior

Heating Efficiencies

Electric co-heating is a technique used to determine the net efficiency of a system consisting of any heating appliance (a furnace, for example) and a house. Several portable, thermostated, and metered electric heaters are distributed throughout the house. First, the furnace is turned off and the electric heaters are used to heat the house. Next the furnace is turned on and cycled manually (for example, four minutes on and sixteen minutes off) for about three hours. Throughout the procedure, the indoor temperature remains constant by appropriate reductions in the heat produced (and power consumed) by the electric heaters. Finally, the furnace is turned off and the electric heaters see the full load again. The indoor temperature is recorded and periodic measurements of the outdoor temperature are taken. Also, the air infiltration rate (through cracks in walls, etc.) is monitored using tracer gas techniques.

The following steps are used to calculate the efficiency: First the heat load, including air infiltration, is determined through use of the electrical heaters. Second, the portion of the heat load supplied by the furnace is found from the difference in the load and the measured electrical power consumed by the heaters. Third, the efficiency is calcualted by dividing the difference by the average furnace fuel consumption rate. Further capabilities of the electric co-heating technique include an evaluation of fireplace efficiency and determination of the fraction of the heating load needed for individual rooms in the dwelling.

Advantages and Limitations. Unlike some methods of efficiency evaluation, electric co-heating takes into account the effect of distribution heat losses (through ducts, etc.); in other words, the efficiency calcualtion includes only the heat that benefits the living space. Also, this method distinguishes between heating efficiency and the house's envelope performance. Electric co-heating, however, is still in the development stage. It is not practical for structures with a large number of rooms because of the difficulty in monitoring the infiltration losses and the power of the heaters.

Physical Inspection—Furnaces and Boilers/Heating Systems

Efficiencies

The current trends toward energy conservation and the subsequent reduction in heating loads in buildings have had an effect on the efficiencies of furnaces and boilers; the older ones tend to be oversized and often yield lower efficiencies. In such cases, a replacement may be desirable. However, several actions can be taken to increase the efficiency of the existing furnace or boiler. The combustion system should be carefully cleaned and checked for leaks. The air should enter the combustion chamber at controlled rates. To obtain optimum efficiencies, the oil nozzle size may need to be reduced or the gas burner may need to be replaced if the system will accommodate a replacement. The air-fuel ratio should be determined and adjusted for proper combustion. The O_2 and CO_2 content can be determined from an Orstat apparatus or other available analyzer such as the nondispersive infrared spectrometer with carbon dioxide or oxygen specific electrodes to determine oxygen content. This, coupled with the stack temperature reading, will provide system data for determining system combustion efficiency. The final decision on system replacement is generally made on the basis of economics—the annual savings that can be expected as a result of purchasing and installing a new system.

Advantages and Limitations. This procedure gives only steady-state performance. In practice, most systems operate far from steady-state; however, this inspection should be undertaken for safety reasons whenever major retrofits of a building are planned. Physical inspection offers the advantage of hands-on experience and an immediate sense of condition of the furnace or boiler and the HVAC system. With physical inspection, there is the need for knowledge and skill gained through experience. The inspector carrying out the condition assessment inspection, should be technically trained. Replacement of furnace of boiler components should be made by qualified personnel only.

Ultrasonic Thickness Gaging—Containment Vessels (Heat Exchangers, Tanks, Pipes, etc.)

Corrosion and Wear

It is possible to use ultrasonic techniques to determine the thickness of materials to obtain an indication of wear and deterioration. The pulse-echo ultrasonic thickness technique is recognized as an accurate method of measuring product thickness when the velocity of ultrasound is known. Ultrasonic transducers have been developed that allow the determination of piping and tubing wall thicknesses (and other forms) without physically penetrating the material. Ultrasonic pulses are generated and the time required for the pulse to be reflected back from the opposite surface is very accurately recorded. By knowing the transmission velocity of the material and using the appropriate electronic circuitry, a digital readout can be obtained which is a direct measure of the thickness of the material. The technique has been used to evaluate large boiler installations for heat exchanger deterioration from corrosion.

Advantages and Limitations. Small transducers are available to determine thickness of closed materials without physically penetrating the surface. By scanning the surface variations, wall thickness can be determined, thus gaining an indication of deterioration. Some distortion results from rough surfaces as the ultrasonic pulse is scattered, with a resulting lack of sharpness in the return echo. As surface curvature increases, the coupling efficiency between transducers and the material to be measured decreases, causing some loss in accuracy. Some materials have special acoustical characteristics and can cause scattering, velocity variation, and attenuation, with a resulting loss in accuracy.

REFERENCE

Test Methods for Existing Mechanical Systems; Selected Methods for Condition Assessment of Structural, HVAC Plumbing and Electrical Systems in Existing Buildings. NBSIR 80-2171; Frank H. Lerchen, James H. Pielert, Thomas K. Faison.

XII. Addenda
Additional Descriptions of
Testing Methods

Testing Methods—Synopsis

Hardness

Liquid-Penetrant Inspection

Limitations
Physical Principles

Magnetic-Particle Inspection

Limitations

Eddy-Current Inspection

Radiographic Inspection

Limitations
Neutron Radiography

Ultrasonic Inspection

Advantages
Disadvantages

Thermal Inspection

Frost Testing

Rapid Identification of Metals and Alloys

Magnetic Properties
Weight
Color
Spark Testing of Ferrous Alloys
Chemical Spot Testing

Residential Case Study

John Yanik, Architect

Subsurface Water Penetration and Damage Caused to a Newly Constructed Rowhouse

Building Pathology in Paradise

Conversation with Dean Elmer Botsai, FAIA, School of Architecture, University of Hawaii at Manoa

ADDITIONAL DESCRIPTIONS OF TESTING METHODS

Testing Methods—Synopsis

Nondestructive and destructive testing can be applied to three phases of construction: during, prior to reconstruction, and existing structures.

Architects and engineers have always used some means of inspection to ensure that the materials and methods they have designed and specified are being used as they have indicated.

As structures and manufactured materials become more complex and more elaborate, more sophisticated tests are developed to test the strength and durability of the materials used. As one of the most common examples of this practice hundreds of concrete cylinders are cast daily, taken to laboratories, cured, and tested. Reports are then made, compared and filed.

Statistical methods are sometimes used to increase the predictability of a limited number of destructive tests. But the value of destructive testing depends on the similarity of the sample destroyed in the test compared to the balance of the material in place in the building. No one can guarantee that this similarity is totally predictable. Nondestructive testing avoids this uncertainity of sampling. It can be applied to all the elements being scrutinized. There is consequently no doubt that units tested have properties similar to those in service, but nondestructive testing is limited, for short of destroying the material by subjecting it to the moment of truth, the maximum real stress the material can withstand, its true strengths and weaknesses will remain in doubt. We simply never know the true ultimate strength of a material until we have destroyed it and what we have gained is absolute certainty about a building element that is no longer usable. Absolute certainty concerning a building in place is therefore unattainable.

The following descriptions of test methods only describe the parameters of current test methods, their skillful use is an art practiced by the diagnostician at least equal in fascination to that of designing the building.

Hardness

The meaning of the term *hardness* varies. It can mean resistance to penetration, resistance to wear, a measure of the flow stress, resistance to scratching, or resistance to cutting. Although these several characteristics may appear to differ they are all related to the plastic flow stress of the material.

The Metals Handbook of the American Society For Metals lists eight methods of testing hardness:

1. Static indentation tests, in which a ball, cone, or pyramid is forced into the surface of the metal tested. The relationship of load to depth of area indentation is the measure of hardness.
2. Rebound tests, in which an object of standard mass and dimensions is bounced from the surface of the workpiece tested. The height of rebound is the measure of hardness.
3. Scratch tests, in which one material is judged capable of scratching another.
4. Plowing tests, in which a blunt element (usually diamond) is moved across the surface of the workpiece being under controlled conditions of load and shape. The width of the groove is the measure of hardness.
5. Damping tests, in which the change in amplitude of a pendulum having a hard pivot which rests on the surface of the workpiece being tested, is the measure of hardness.
6. Cutting tests, in which a sharp tool of given shape removes a chip of standard dimensions from the surface of the workpiece tested.
7. Abrasion tests, in which a workpiece is loaded against a rotating disk and the rate of wear is the measure of hardness.
8. Erosion tests, in which sand or other granular abrasives are impinged upon the surface of the workpiece tested under standard conditions. The loss of material in a given time is the measure of hardness. Hardness of grinding wheels is measured by this method.

Liquid-Penetrant Inspection

This is a descendent of the old oil-and-whiting prodess. It relies on the ability of the penetrant to seep, through surface tension and capillary action, into a surface discontinuity. The method is limited to surface defects or subsurface defects with surface openings. Penetrant testing is best adapted to inspecting all types of surface cracks, porosity, laminations, lack of bond at exposed edges of joined materials, and leaks in tubing, tanks, welds, and similar items.

It is a method for finding discontinuities open to the surface of solid and essentially nonporous materials. Indications of flaws can be found regardless of the size, configuration, internal structure, or chemical composition of the workpiece inspected and regardless of flaw orientation. Liquid penetrants can seep into and be drawn into various types of minute surface openings, reportedly as fine as 4 micro-in. in width, by capillary action. The process is well suited for the detection of all types of surface cracks, laps, porosity, shrinkage areas, laminations, and similar discontinuities. It is used extensively for the inspection of wrought and cast products of both ferrous and nonferrous metals, ceramics, plastics, and glass objects.

In practice, the liquid-penetrant process is relatively simple. Equipment is generally less costly than that used for most other nondestructive inspection methods. The liquid penetrant may contain a fluorescent dye. It is applied to the surface of the test item by dipping, spraying, or wiping and is allowed to penetrate the surface of the material. Excess penetrant on the surface is washed off, and an absorbent, light-colored powder, called a developer, is applied to the surface. The developer draws the penetrant out of the surface openings, and the penetrant stains an area much wider than the corresponding surface law. The flaws are then examined under ultraviolet light.

Limitations

The major limitation is that this method only detects imperfections open to the surface. Other methods must be used to detect subsurface defects or discontinuities.

Physical Principles

Inspection depends mainly on a liquid's effectively wetting the surface of a solid workpiece, flowing over the surface to form a continuous and reasonably uniform coating, and then migrating into cavities that are open to the surface. The cavities are usually exceedingly small, often invisible to the unaided eye. The ability of a given liquid to flow over a surface and enter surface cavities depends principally on the following factors:

1. Cleanliness of the surface
2. Configuration of the cavity
3. Size of the cavity
4. Surface tension of the liquid
5. Ability of the liquid to wet the surface.

Magnetic-Particle Inspection

Magnetic-particle inspection is limited to ferromagnetic materials that can be intensely magnetized. It is not applicable to materials such as aluminum, bronze, brass, and copper.

The parts to be inspected are magnetized with a permanent magnet, or more often with an electromagnetic field. Small magnetic particles, which may be red or black or may contain fluorescent pigment, are applied to the surface of the test item. Surface defects, or in some cases subsurface defects, create magnetic poles around the outline of the discontinuity, and the colored particles tend to collect in this region.

This is an excellent method for finding cracks in ferrous materials, which may be caused by fatigue, heat treating and quenching, or grinding. In construction it is used to discover incomplete weld penetrations. There is no practical limit to the size and configuration of the test item.

It is a method for locating surface and subsurface discontinuities in ferromagnetic materials. It depends for its operation on the fact that when the material or part under test is magnetized, magnetic discontinuities that lie in a direction generally transverse to the direction of the magnetic field will cause a leakage field to be formed at and above the surface of the part. The presence of this leakage field, and therefore the presence of the discontinuity, is detected by the use of finely divided ferromagnetic particles applied over the surface, some of the particles being gathered and held in the leakage field. This magnetically held collection of particles forms an outline of the discontinuity and generally indicates its location, size, shape, and extent. Magnetic particles are applied over a surface as dry particles, or as wet particles in a liquid carrier such as water or oil.

Limitations

There are certain drawbacks to magnetic-particle inspection:

1. Thin coatings of paint and other nonmagnetic coverings, such as plating, adversely affect sensitivity of the inspection.
2. The method works only on ferromagnetic materials.
3. Demagnetization following inspection is often necessary.
4. Cleaning to remove remnants of the magnetic particles clinging to the surface may sometimes be required.
5. Exceedingly large currents are sometimes required for very large parts.
6. Care to avoid local heating and burning of finished parts or surfaces at the points of electrical contact is necessary.
7. Although indications are easily seen, experience and skill in interpreting their significance is required.

Eddy-Current Inspection

This method is based on the principles of electromagnetic induction and is used to identify or differentiate between a wide variety of physical, structural and metallurgical conditions in electrically conductive ferromagnetic and nonferromagnetic metals and metal parts. Eddy-current inspection can be used:

1. To measure or identify such conditions and properties as electrical conductivity, magnetic permeability, grain size, heat treatment condition, hardness and physical dimensions
2. To detect seams, laps, cracks, voids, and inclusions
3. To sort dissimilar metals and detect differences in their composition, microstructure, and other properties
4. To measure the thickness of nonconductive coating on a conductive metal or the thickness of a nonmagnetic metal coating on a magnetic metal.

Radiographic Inspection

Radiographic inspection makes a permanent photographic record of the internal condition of the material being tested. Either X-rays from an electronic radiation source, or gamma rays emitted by radioactive isotopes may be used. (A radiograph is produced when the rays pass through a test sample onto a film.)

The short-wavelength characteristic of both x-rays and gamma rays permits the beam to penetrate opaque materials. The composition, thickness, and density of materials affect their absorption of x-rays or gamma rays.

With a homogeneous test specimen, the darkness, or film density, of the exposed and processed film will be uniform. Internal discontinuities and flaws, as well as intentional holes, appear darker than the surrounding area in a well-defined radiograph. If the material being examined contains a foreign body with a density different from the base material, a corresponding difference in darkness will appear in the developed film.

The overall quality of a radiographic inspection can be checked with a penetrameter. A penetrameter is a thin plaque of metal containing three holes of graded diameters, usually $2t$, $3t$, and $4t$, where t is the thickness of the penetrameter itself. This thickness is related to the sample being inspected, and is usually 2% of the sample.

The penetrameter is placed on the source side of a specimen before it is exposed to radiography. The radiography may be considered satisfactory if the penetrameter and its holes show clearly on the radiograph.

Portable-tank x-ray machines enable testers to make filmed inspections where electric power is available. For areas without electric power, or where it is considered desirable to make a full 360° peripheral inspection of a welded joint in a pipe or similar structure, a radioactive-isotope gamma ray source is used.

Radioactive isotopes are used for this application because they radiate in all directions, whereas the output from conventional x-ray sources is essentially straight line. By wrapping a strip of film around the circumference of a pipe and inserting a suitably shielded radioactive isotope into the pipe, the whole perimeter can be radiographed.

Radiography is used for inspection of components and assemblies based on differential absorption of penetrating radiation (either electromagnetic radiation or very-short-wavelength particulate radiation) by the part or test piece being inspected.

Because of the differences in density and variations in thickness of the part, or differences in absorption characteristics caused by variations in composition, different portions of the test piece absorb different amounts of penetrating radiation.

The term *radiography* usually implies a radiographic process that produces a permanent image or film (conventional radiography) or paper (paper radiography or xeroradiography) although in a broad sense it refers to all forms of radiographic inspection.

When inspection involves viewing of a real-time image on a fluorescent screen or image-intensifier, the radiographic process is termed *fluoroscopy*. When electronic instruments are used to measure the intensity of radiation, the process is termed *radiation gaging*. All these terms are used mainly a connection with inspection that involves penetrating electromagnetic radiation in the form of x-rays or gamma rays. *Neutron radiography* refers to radiographic inspection using a stream of neutrons rather than electromagnetic radiation.

The use of x-rays or gamma rays can be hazardous. Such radiation can produce biological damage to body tissues; blood cells, the lenses of the eyes, and certain internal organs are particularly sensitive to radiation. Consequently, strict control of human exposure to radiation is mandatory.

Both ferrous and nonferrous alloys can be radiographed, as can nonmetallic materials and composites.

Limitations

Radiography is expensive. Relatively large capital costs and space allocations are required for radiographic laboratory, fluoroscopic inspection station, or image intensifier equipment. When portable x-ray or gamma-ray sources are used, capital costs can be relatively low. Space is only required for film processing and interpretation. Operating costs can be high; as much as 60% of the total inspection time is spent in setting up for radiography. With fluoroscopy, operating costs usually are much lower, because setup times are shorter and there are no extra costs for processing or interpretation of film.

Neutron Radiography

This radiographic method uses a specific type of particulate radiation, called neutrons, to form a radiographic image of a test piece. The geometric principles of shadow formation, variation of attenuation with test-piece thickness, and many other factors that govern the exposure and processing of a neutron radiograph are similar to those for radiography using x-rays or gamma rays.

Neutron radiography is not accomplished by direct imaging on film, because neutrons do not expose x-ray emulsions efficiently. In one form of neutron radiography, the beam of neutrons impinges on a screen made of a material such as dysprosium or indium, which absorbs the neutrons and becomes radioactive, decaying with a short half-life. In this method, the screen alone is exposed in the neutron beam, then immediately placed in contact with film to expose it by autoradiography. In another common form of imaging, a screen that immediately emits secondary radiation is used with film directly in the neutron beam.

Ultrasonic Inspection

The principle of ultrasonic inspection is based on the ability of high-frequency sound waves (those above the upper limit of the audible range) to travel with little loss through homogenous material, except when intercepted and reflected by discontinuities in the interior of the material.

The basic ultrasonic system consists of a signal generator, a transducer to reverse the process after receiving a transmitted or reflected signal, and a means of displaying the signal.

There are two types of ultrasonic tests. In one, a beam of ultrasonic energy directed into a test specimen is transmitted through the material and is picked up by a receiver on the opposite side. This is called the *through-transmission method*. The other method, called the *pulse-echo method*, reflects the energy back to the same side of the material as the transmitter.

In addition to detecting flaws, ultrasonic testing is suitable for determining the thickness of material that is inaccessible on two sides. Engineers use it to measure corrosion in tanks and pipes.

Among the advantages of ultrasonic testing are: high sensitivity that permits the detection of minute subsurface defects; great penetration power that allows very thick sections to be examined; accuracy in determining flaw position and size; fast response to rapid and automated inspection; and ability to work with only one available surface (pulse-echo method).

However, unfavorable sample geometry, such as size, contour, complexity, defect orientation with relation to the surface, and normal inclusions, such as aggregate in concrete, limit the application of the ultrasonic method.

Ultrasonic inspection is one of the most widely used methods of nondestructive inspection. Its primary application in the inspection of metals is the detection and characterization of internal flaws; it is also used to detect surface flaws, to define bond characteristics, to measure thickness and extent of corrosion, and (much less frequently) to determine physical properties, structure, grain size and elastic constants.

Advantages

1. Superior penetrating power, which allows the detection of flaws deep in the part.
2. High sensitivity, permitting the detection of extremely small flaws.
3. Greater accuracy than other nondestructive methods in determining the position of internal flaws, estimating their size and characterizing their orientation, shape, and nature.
4. Only one surface need be accessible.
5. Operation is electronic, which provides almost instantaneous indications of flaws.
6. Is not hazardous to operations or to nearby personnel.
7. Portability.

Disadvantages

1. Manual operation requires careful attention by experienced technicians.
2. Extensive technical knowledge is required for the development of inspection procedures.
3. Parts that are rough, irregular in shape, very small or thin, or not homogeneous are difficult to inspect.
4. Discontinuities that are present in a shallow layer immediately beneath the surface may not be detectable.

Thermal Inspection

This is inspection by all methods in which heat-sensing devices or substances are used to detect irregular temperatures. There are several methods of thermal inspection and many types of temperature-measuring devices and substances.

Thermography, which is the mapping of isotherms, or contours of equal temperature, over a test surface and *thermometry,* which is the measurement of temperature, are discussed here.

Techniques are separated into two categories: (a) direct contact, in which a thermally sensitive device or material is placed in physical and thermal contact with the test piece; and (b) noncontact techniques that depend on thermally generated electromagnetic energy radiated from the test piece. At moderate temperatures, this energy is predominantly infrared. Therefore, infrared testing is an important branch of noncontact thermal testing.

Workpieces to be inspected by thermal techniques are considered either passive or active. Passive pieces are those that are artificially heated or cooled during inspection to obtain a thermal profile. The presence of an imperfection is indicated by an abnormal temperature in the vicinity of the imperfection. Usually, the greater the impedance of the material to heat flow the more readily the imperfection can be found. This is true because materials that have high resistance to heat flow change temperature more slowly than those with low resistance, and therefore equilibrate more slowly when tested under fixed thermal input. Because of this property, there is a temperature gradient around the imperfection; the closer it is to the surface being inspected, the greater the temperature difference.

Frost Testing

This technique limited application but it can be used as a quick method for determining the presence of gross flaws. In frost testing the surface of the workpiece is chilled to a temperature below the freezing point of water so that a frosted surface is obtained. Visual observation of the frost pattern as it melts indicates the areas that are relatively high and low in thermal conductivity, and thus reveals the location of flaws. Chilling can be accomplished by spraying the surface with a refrigerant gas, such as Freon or carbon dioxide, or by cold-soaking the workpiece in a low-temperature chamber and then exposing the workpiece at room temperature.

Rapid Identification of Metals and Alloys

Common methods of rapid identification of metals include techniques involving the magnetic properties, weight, and color of the metal. Spark testing and chemical spot testing also are used.

Magnetic Properties

Applying a hand magnet to a metal is the simplest method of distinguishing between ferromagnetic and nonferromagnetic metals. The attraction on the hand magnet to the metal will be strongly magnetic, slightly magnetic, or nonmagnetic. Ferromagnetic metals, particularly alloys of iron, cobalt, and nickel, are attracted or repulsed strongly.

Weight

Metals can be separated into groups on the basis of their weight (specific gravity).

Color

Some metals and alloys can be identified by their color. Gold and silver can be identified by their distinctive colors. Depleted uranium is dark yellow; the oxidized surface may be brown. The coppers are reddish or red-brown, bronzes are dark yellow, and brasses are light yellow. Colors of the copper alloys can vary depending on the alloying elements and content. All other metals and alloys in the unoxidized condition are white or grayish white.

Spark Testing of Ferrous Alloys

This method is used for the classification of ferrous alloys according to their chemical compositions, by visual examination of the spark pattern or stream that is thrown off when the alloys are held against a grinding wheel rotating at high speed. The test is a fast and economical method of separating alloys of different compositions. Experienced operators can use this method for identification of a number of ferrous alloys with reasonable accuracy.

Spark testing is not intended as a subsitute for chemical analysis. Each spark stream from a number of parts supposedly of the same composition should have the same spark characteristics. If one or more of the parts exhibits a spark stream having spark characteristics that are different from the other parts being tested, a different composition is indicated. In such instances, chemical or spectrochemical analysis can be used for positive identification of the unknown composition.

Chemical Spot Testing

Chemical spot tests make possible a quick field identification of metals by means of chemical color reactions, taking place solely in one spot on a sample, on filter paper, or on a spot plate. Tests can be made at room temperature and without the need for any apparatus other than a simple test kit. The tests should be used as a sorting or screening procedure and not as a substitute for quantitative chemical analysis.

REFERENCES

American Society for Metals, *Metals Handbook,* Volume 11, *Nondestructive Inspection and Quality Control,* Metas Park, Ohio, 44073.

Yoder, Carl B. and Dadson, E. J., of the United States Testing Company, Inc., Hoboken, N.J., "Nondestructive Testing in Building Construction," *P/A,* Sept., 1966.

RESIDENTIAL CASE STUDY

John Yanik, Architect

Subsurface Water Penetration and Damage Caused to a Newly Constructed Rowhouse

In September, 1977, the distraught owner of a newly constructed rowhouse located in Alexandria, Virginia, telephoned to ask if I would make an inspection to determine why he was getting water in his basement family room. The owner was quite upset because he had moved into the house right after construction was completed and had reported a number of defects to the builder, but none of them had been corrected. The water problem, and the damage it was causing to floor tile and wall paneling in this room, was the last straw as far as the owner was concerned. I agreed to take a look at it.

The two-story house with a finished basement was the last in a row at the end of a pedestrian way leading from a common parking area to the house entrance. The house was downslope and approximately 8 ft lower in elevation than the first house in the row. The family room was located at the real of the basement and water damage was clearly visible along the rear wall of the room. A garden hose had been allowed to run at a medium flow for one hour to soak the ground in the outside patio area above prior to my arrival. The owner had also removed several sections of paneling to expose the inside face of the wall and had taken up the carpeting to expose the floor tile. Many of these tiles had buckled from dampness and I was able to observe the water running down the face of the exposed masonry in rivulets. Wood furring and sill plates were rotted in several areas. Possible causes for the water penetration were the following:

1. Lack of a positive slope in the patio above which would direct the surface water *away* from the house. The water was ponding in a low spot above and near the basement wall, and the drainage was further aggravated by a masonry patio wall at right angles to the end of the house. This wall, which had no drain holes at grade, tended to trap the water and prevent it from continuing its run downslope.
2. A second cause would have to be inadequate or defective parging on the exterior face of the basement wall. This could not be visually inspected without excavating.
3. A third contributing cause could be the lack of a properly functioning footing drain tile. No outlet for the drain could be found in the slope beyond the end of the house. Again, without excavating, there was no way to be sure that a footing drain had been provided; or, if it had been provided, whether or not it was connected to a storm drain.

The report, subsequently prepared, called for the excavation of the area adjacent to the wall down to the footing, the providing of a properly functioning footing drain, the reparging of the entire below-grade face of the exterior wall, backfilling and compacting the patio area with clay-free soil, regrading, and the provision of drain holes at the base of the patio wall. This report, along with other documentation and photos provided by the owner, was to be used in a demand for arbitration filed by the owner under the terms of the owner's HOW (Home Owner's Warranty) agreement.

Before the date set for the arbitration arrived, there was another frantic call from the owner. Water was now appearing in several other areas of the basement along the end wall and front wall. It was clear, therefore, that the owner's house was under more severe water pressure than first suspected. An additional complaint had also developed. The owner pointed out several cracks which had appeared on the second floor at the juncture of wall and ceiling. There was not time to investigate this fully before the arbitration; but it was decided to do so immediately afterward. For the moment, an addendum to the first report was issued citing more water penetration in other areas of the basement. The remedy called for in the first report remained; but the addendum called for the added installation of a internal subfloor drain system on three sides of the basement with gravity flow to an outlet downslope of the house.

The first arbitration ended successively for the owner and he was awarded the remedy called for in the report *or* the payment of $11,250.00 by the builder. This was the largest monetary arbitration award against a Virginia builder to date and was reported in the Washington Post.

I now made a more detailed inspection of the open joints at the intersection of the wall and ceiling at the second floor and climbed up into the shallow attic space above. The largest separation had occurred along line (d) (see sketch plan of second floor) at a bedroom wall located approximately midway between the front and the rear of the house. The roof was framed with prefabricated 2 × 4 wood trusses spanning over this wall and bearing on the front and real walls of the house. Between the bottom of the truss and the top of the 2 × 4 plate for the wall below, I measured a gap varying from ¾ in. to ⅞ in. Since the ceiling gypsum wallboard was nailed to the underside of the truss, the separation noted in the ceiling below could be accounted for by only one of two possible occurrences: either the bottom chords of the trusses had shortened their length through shrinkage and drawn upward "arch" fashion *or* the wall had settled. I noted the location of this and all other cracks in the drywall at the second floor on a sketch diagram. They were as follows:

Cracks and separations in drywall (see attached floor plan sketches and diagrammatic section through house, drawings, no. 1, 2, and 4):

(a) Master Bedroom: crack in wall above right door jamb (door to hall); crack above left door jamb (door to Master Bath). Both doors will not close by ³⁄₁₆ in., indicating that some settlement has taken place along the center of this wall.

(b) Master Bath: cracks observed between ceiling and wall in the two locations indicated on sketch.

(c) Bathroom off Hall: a substantial seperation has occurred between ceiling and wall above the tub.

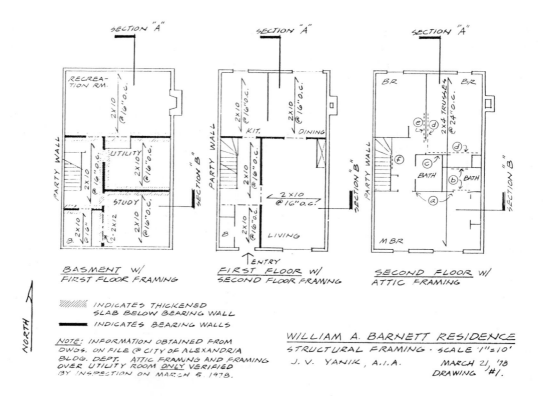

BASEMENT w/
FIRST FLOOR FRAMING

FIRST FLOOR w/
SECOND FLOOR FRAMING

SECOND FLOOR w/
ATTIC FRAMING

////// INDICATES THICKENED
SLAB BELOW BEARING WALL

▬▬ INDICATES BEARING WALLS

NOTE: INFORMATION OBTAINED FROM
DWGS. ON FILE @ CITY OF ALEXANDRIA
BLDG. DEPT. ATTIC FRAMING AND FRAMING
OVER UTILITY ROOM ONLY VERIFIED
BY INSPECTION ON MARCH 5 1978.

WILLIAM A. BARNETT RESIDENCE
STRUCTURAL FRAMING · SCALE 1"=10'
J. V. YANIK, A.I.A.   MARCH 21, '78
                           DRAWING #1.

(d) Right Rear Bedroom: a very pronounced fold has occurred along the ceiling drywall joint 2½ in. from the face of the closet wall. Some folding is also visible between the ceiling and wall at the sidewall near the latch side of the Bedroom door.

(e) Left Rear Bedroom: a similar fold is occurring at the sidewall ceiling near the hinge side of this door.

(f) Hall: a crack has appeared between ceiling and wall at stair.

That these cracks could have been produced by a shrinkage of the truss members appeared highly unlikely. The air-conditioning system for the house provided for automatic humidification during the heating season so that, even if the trusses had been fabricated from improperly kiln-dried lumber, there would be sufficient moisture present to prevent excessive shrinkage. A check of shrinkage characteristics for wood also indicated that, even under the worst circumstances, a gap of ⅛–⅛ in. might have produced, but not the ¾–⅞ in. measured.

Next, since the owner did not have a set of construction drawings for the house, I inspected a set which was on file at the Alexandria Building Department. I traced enough information from these drawings to enable me to prepare accurate floor plan diagrams showing the structural framing for the first and second floors, sizes of members, direction of span, location of bearing walls, and type of foundations. I was also able to find the site grading and drainage drawings for

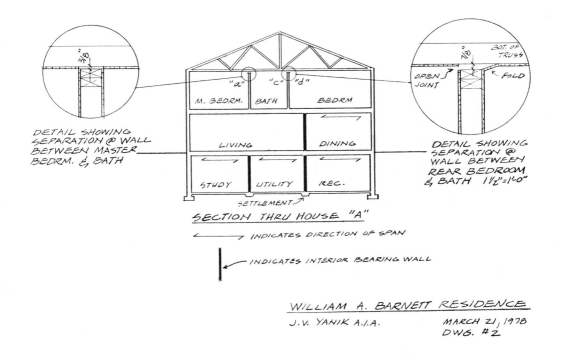

DETAIL SHOWING
SEPARATION @ WALL
BETWEEN MASTER
BEDRM. & BATH

DETAIL SHOWING
SEPARATION @
WALL BETWEEN
REAR BEDROOM
& BATH 1½"=1'-0"

SECTION THRU HOUSE "A"

⟵——⟶ INDICATES DIRECTION OF SPAN

| ⟵ INDICATES INTERIOR BEARING WALL

WILLIAM A. BARNETT RESIDENCE
J. V. YANIK A.I.A.   MARCH 21, 1978
                           DWG. #2

the subdivision. By locating my client's house on these drawings, I made the interesting discovery that his house had been constructed on *6–8 ft of compacted fill over an existing stream* (see drawing No. 3). I was able to obtain the name of the design engineer for the consulting firm which prepared site grading and drainage drawings and arranged to meet with him. He informed me that they had inspected the site recently and had found that an 8-in. perforated "underdrain" was missing or unconnected to the catch-basin southeast of my client's house. Since that 8-in. drain was to carry the flow from three other 6-in. lines in that area of the site, the engineers were quite concerned and had written to the builder asking him to investigate this and determine why this was so. What this meant, is that the system to carry away subsurface water in the area where my client's house was located was inoperative!

All the evidence now pointed to the settlement of several interior bearing walls of the house as the cause of the cracks in question. The analysis described above was spelled out in a second report to the owner along with the following conclusions:

A. Structural framing design appears to be adequate and consistent with good practice. All available evidence indicates that the major plaster cracks visible at the second floor of the house are being caused by the *settlement* of two *interior* bearing walls (see basement plan):

   1. The east-west bearing wall between the recreation room and the utility room

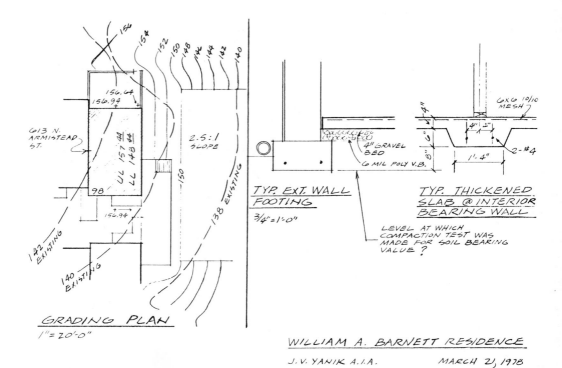

GRADING PLAN
1" = 20'-0"

613 N. ARMISTEAD ST.

2.5:1 SLOPE

156.64
156.94

UL 157 ↓↓
LL 148 ↓↓

98

156.94

142 EXISTING

140 EXISTING

138 EXISTING

TYP. EXT. WALL FOOTING
3/4" = 1'-0"

4" GRAVEL BED
6 MIL POLY V.B.

LEVEL AT WHICH COMPACTION TEST WAS MADE FOR SOIL BEARING VALUE ?

TYP. THICKENED SLAB @ INTERIOR BEARING WALL

6X6 10/10 MESH

2 - #4

1'-4"

WILLIAM A. BARNETT RESIDENCE

J. V. YANIK A.I.A.   MARCH 21, 1978
                            DWG. #3

NOTE: INFORMATION OBTAINED FROM DWGS. ON FILE @ CITY OF ALEXANDRIA BLDG. DEPT.

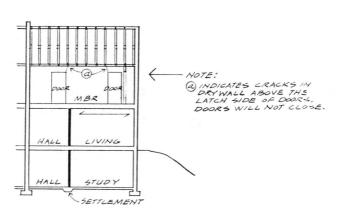

DOOR   @   DOOR
MBR

HALL   LIVING

HALL   STUDY

SETTLEMENT

NOTE:
@ INDICATES CRACKS IN DRYWALL ABOVE THE LATCH SIDE OF DOORS. DOORS WILL NOT CLOSE.

SECTION THRU HOUSE "B"
⟶ INDICATES DIRECTION OF SPAN

| INDICATES INTERIOR BEARING WALL

WILLIAM A. BARNETT RESIDENCE

J. V. YANIK A.I.A.   MARCH 21, 1978
                            DWG. #4

2. The north-south bearing wall between hall/bathroom and utility/study. As these walls began to settle, cracks were opened up at the intersection of walls and second floor ceiling because the ceiling wallboard is nailed to the bottom chord of the trusses, and trusses are *exterior* wall bearing.

B. *Exterior* bearing walls do not appear to have settled to any appreciable degree.

C. The exact cause of internal bearing wall settlement is unknown. Inadequate soil compaction *may* have been contributing cause for the final 8-in. height difference between the bottom of the exterior footing and the underside of the thickened slab. Subsurface water pressure may very likely have been a contributory cause, since it could have produced some additional compaction through the rearrangement and resettling of fines in the soil, or may have begun an actual erosion of soil under the thickened slabs.

D. Any remedy to stop bearing wall settlement, should be related to a fully adequate under-slab tile drainage system.

E. While there is no immediate structural danger caused by the settlement to date, continued settlement would have a vital effect on the use of the house for residential purposes.

Since a soil bearing test was required only for the exterior wall footings, and an additional 8 in. of fill would have been added to form a base for the thickened slabs, a careless contractor might well have neglected to compact the last 8 in.

As a remedy, the second report called for an interior drain below the slab at the end wall of the house in addition to the drains called for in the addendum to the first report, and for a properly engineered underpinning system to be installed under the two interior bearing walls.

The owner made a formal application for a second arbitration hearing and it was granted. Before the date set for the hearing arrived, I received an excited call from him. The builder had offered to buy back the house and they had agreed on a price which was more than satisfactory to the owner.

This owner, who was in danger of being victimized by defective and careless building practices, was able to obtain satisfaction because he was willing to expend the time, energy, and money to document his case. Not every owner is willing or able to do so.

Footnote

About a year later, after my client had moved into an older house in a well-established neighborhood, I received a telephone call from an attorney. The attorney said that he represented Dr. ———— who had purchased a house at ———— — Street, Alexandria. The Doctor was very upset to find puddles of water forming in his basement and several cracks had appeared at the ceiling of one of the upstairs bedrooms. The attorney had been given my name by Mr. ————. "Do you have any idea what the problem might be? The builder who sold the house to my client has been trying to minimize the situation . . ." etc., etc.

The address given was that of the same house I had investigated a year earlier and have described above. I can only surmise that the builder did a certain amount of cosmetic work to conceal the problem and then found another unsuspecting buyer.

BUILDING PATHOLOGY IN PARADISE

Conversation with Dean Elmer Botsai, FAIA, School of Architecture, University of Hawaii at Manoa

Engineering and the use of the computer has become a refined science that can solve problems to the nth degree. But the entire premise, the hypothesis of enginerring and computer investigation rests on capricious and arbitrary guesses. If the fundamental premise is fallacious then the entire project, the edifice of thought on which it is based, is worthless.

Most building failures, or the great majority of them, will be found in the building's connections. Failure occurs at the contact points where materials meet, where similar and dissimilar materials are joined and where sealants are stuck to other materials. Failure occurs in the joints. We seldom or never hear of a member failing because of an inability to withstand the forces it was designed to neutralize.

When surveying a building look to the connections. They usually cannot be seen. You must, therefore, search for signs and indications of their condition. Your senses must be turned to detect signs of things that do not belong where they are. Develop a sense of smell, a sense of feel, a sense of sound for a sick building. Feel the building's vibration. All of your senses must come into play, for each must be used.

As an example of the use of this intuitive perception, when I was called to inspect a building for water leaks I saw the surface sign of infirmity in spalled concrete. Engineers had made an inspection previously and found nothing amiss. I asked them to go back and examine the joint connections between poured in place and precast concrete units. They did so and their conclusion was the entire building was hazardous. The engineers were competent, but their inspection was conducted assuming nothing wrong. They inspected a healthy building and examined it using standard procedures. When they returned a second time they examined the building with an altered frame of mind. They studied the building's history in the design and shop drawings as well. The structure, a parking garage, is now being repaired.

In another instance I was retained to inspect a complex of buildings suffering from water damage. When I first walked through them my instincts warned me that something was terribly wrong. They had been beautifully sited, but there was water in the basements and water stains everywhere. There was a smell. The sound of the plaster when I tapped was wrong. We applied surgery to the exterior skin and the resulting diagnosis was extensive, tragic damage due to water and termite infestation. Damage repair amounted to 3½ million dollars. The cause was design midjudgment and slovenly construction.

Damage due to water infiltration and termite infestation, Hawaii. (Photo by F. Wilson)

Corrective measures (see above). (Photo by F. Wilson)

(Photo by F. Wilson)

The building did not sound, look, smell, or feel right. A porch fell and people were injured. It was actually fortunate that no one was killed. Damage was so extensive we found a 6 × 16 inch beam decayed entirely through. There were 87 units with 60 decks and every deck failed. Such massive, consistent, failure made it impossible not to blame design judgment and construction performance.

Design in Hawaii, because of the benign climate, appears to be deceptively simple. In reality the combination of water, termites, and moisture make this an extremely difficult place to build.

In another instance when called to examine water infiltration I had a similar premonition of building illness. Three floors of wood construction rested on a post-tensioned concrete slab. When I walked on the slab it felt dead. I checked it and found large cracks in the underside. It was no longer a taut, post-tensioned slab but a flaccid catenary. We jacked it to level. The cracks then passed entirely through from top to bottom. We filled the cracks with epoxy and removed the jacks. The slab has regained some of its original compressive strength.

I am not comfortable with the practice of post-tensioning. Few craftsmen are capable of exercising sufficient skill to make it work. It is usually cheaper and surer to pour concrete in place. A major problem of our time, that no one seems ready or willing to face is, how will these post-tensioned buildings be disassembled?

There are serious gaps in our knowledge of material performance. We should, for example, know more of high-strength steel corrosion. We know that radar domes on jets experience an analogous type of failure yet have not availed ourselves of this knowledge.

In Hawaii the military are studying a serious problem of airplane corrosion. We simply do not know enough about the materials we use and do not seem to want to learn from others with experience that might help us.

We pressure treat Douglas fir, but do not treat it correctly. We should either change our materials or alter our method of pressure treatment. A chemical change is needed.

The Douglas fir plywood panel is a classic example of irresponsible material specification and use. The panel specification applies to the face veneer only. The specifications benignly state that the core can be fabricated of a comparable wood in strength. It can be manufactured of alder or pine which have almost no decay resistance at all. The inner plies are therefore more susceptible to decay. The exterior grade plywood remains undamaged although the body of the panel behind it has entirely disintegrated.

We are not told clearly that clay products expand when wet and do not shrink again when they dry. Brick veneers walk off the face of buildings. The expansion properties of clay paving tile are quite serious and accumulate to a great deal of movement.

When designing foundations we are intelligent enough to call clay an expansive soil yet do not give it proper respect in clay masonry units. I and most architects find the natural glaze tone of clay attractive. It is a beautiful finish yet it becomes unstable when wet.

Thinset latex mortar is another danger; its setting time may be 1½ minutes. If tile is not imbedded immediately there is no adhesion. This is not said on the label. The designer and the workmen do not know that the adhesive sets as it is spread.

ASTM standards are somewhat peculiar. They specify ⅛ in. allowable tolerance in manufacture, which adds to ¼ in. in variation from a hypothetical true line. A door that is this amount out of plumb is almost impossible to hang.

We find roofing felt on the market with glass binders that are water soluble. It is a superior roofing material as long as it is not exposed to water. We find the flash point of roofing asphalts equal in temperature to the heat required to apply them, and we accept kraft paper behind plaster walls.

Glass roofing felt has a memory. The manufacturer will not tell the designer this. On a hot day asphalt softens and the glass felt tries to regain its original shape. When applying glass felt with mechanical spreaders, the spreaders leave tracks. The asphalt must cool and harden. If this is not allowed, then the mechanical spreaders will leave roadways in the felt. This fact, remember, was not told to you by a manufacturer, but by me.

If buildings sheathed in granite and marble use ferrous metal connectors, oxidation will push the granite off the face. Cadmium plated fasteners with aluminum spells disaster. Manufacturers do not care what the specifications say, aluminum and cadmium are combined.

There also does not seem to be much study undertaken on how an aluminum wall will move under thermal load. Given 150 degree change over a year's time a 10-ft piece of aluminum will move ⅜ in.

Concrete responds to moisture, like wood. If moisture cannot leave from both sides, concrete, like wood, will curl. It takes moisture and the wet surface expands and curls around the dry one.

I inspected some six-foot-square panels poured on top of a flat slab in a checkerboard pattern. All were curled at the edges. They had lifted slightly, water seeped in around the edges. Then with moisture on one side and the hot sun on the other they went wild.

Plastic cements are an interesting material, for their shrinkage is twice that of Portland cement. Some brands are fine, others not.

We find insulated glass with a five-year life expectancy, after which a good deal of condensation can be expected.

I would tell designers that one does not have to look carefully but to "know." A bit of common sense is never amiss. A natural inquisitiveness is part of an interest in how things work and is important. We must not be satisfied with how things are done, but find how to do them right.

The architectural profession seriously lacks technical competence. A direct correlation can be drawn between the increase in the number of liability and insurance claims and the number of specification writers who never get into the field. These are manufacturer's whores. No, they are not even prostitutes, for they are not paid or kissed.

Designers should be concerned.

Reference Sources

AMERICAN CONCRETE INSTITUTE (ACI)
P.O. Box 19150
Redford Station
Detroit, MI 48219

AMERICAN PLYWOOD ASSOCIATION
P.O. Box 11700
Tacoma, WA 98411

AMERICAN SOCIETY FOR METALS
Metals Park
Ohio, 44073

AMERICAN SOCIETY FOR TESTING AND MATERIALS (ASTM)
1916 Race Street
Philadelphia, PA 19103

AMERICAN SOCIETY OF HEATING, REFRIGERATION AND AIR-CONDITIONING ENGINEERS INC. (ASHRAE)
345 E. 47th Street
New York,
N.Y. 10017

BRICK INSTITUTE OF AMERICA (BIA)
1750 Old Meadow Road
McLean, VA 22102

FOREST PRODUCTS LABORATORY (FPL) OR (USDA FPL)
U.S Dept. of Agriculture
P.O. Box 5130
Madison, WI 53705

METALS ENGINEERING INSTITUTE
Metals Park
Ohio, 44073

NATIONAL BUREAU OF STANDARDS (NBS)
Center for Building Technology
Washington, D.C., 20234

NATIONAL FOREST PRODUCTS ASSOCIATION (NFPA)
1619 Massachusetts Avenue, N.W.
Washington, D.C. 20036

STRUCTURAL CLAY PRODUCTS ASSOCIATION
changed to Brick Institute of America (BIA)

U.S. DEPT. OF HOUSING AND URBAN DEVELOPMENT (HUD)
451 Seventh Street, S.W.
Washington, D.C., 20411

WOOD READINGS

AMERICAN SOCIETY FOR TESTING AND MATERIALS (ASTM)

"Standard Methods for Establishing Clear Wood Strength Values," ASTM D 2555, 1976.
"Standard Methods for Establishing Structural Grades and Related Allowable Properties for Visually Graded Lumber," ASTM D 245, 1974.
"Standard Test Methods for Moisture Content of Wood," ASTM D 2016, 1974

Other Readings—Wood

Berger, H., "Radiography as a Tool of Nondestructive Testing," *Forest Products Journal*, July 1964.
Duff, J. L., "A Probe for Accurate Determination of Moisture Content of Wood Products in Use," U. S. Forest Service Research Note, FPL 0142, 1966.
Elvery, R. H., "Strength Assessment of Timber for Glued Laminated Beams," Symposium of Nondestructive Testing of Concrete and Timber, London, 1969.

Gerhards, C. C., "Stress Wave Speed and MOE of Sweetgum, Ranging from 150 to 15 percent MC," Forest Products Lab, USDA, *FPL Journal,* 1974.

Hart, D., "X-Ray Inspection of Historic Structures: An Aid to Dating and Structural Analysis," TECHNOLOGY AND CONSERVATION, Summer 1977.

James, W. L., "Electric Moisture Meters for Wood," USDA, Forest Service General Tech. Report FPL-6, 1975.

Mackay, J. F. G., "Effect of Moisture Gradients on the Accuracy of Power Loss Moisture Meters," FPL (no year).

McDonald, K. A., "Ultrasonic Location of Defects in Softwood Lumber," USDA, Technical Article in the *Timber Trades Journal,* January 1973.

McDonald, K. A., "Lumber Defect Detection by Ultrasonics," FPL, USDA, Research Paper FPL 311, 1978.

"National Design Specification for Wood Construction," National Forest Products Association, 1977.

"Pilodyn, A Non-destructive Wood Tester," manufacturer's literature, Carl Bechgaard, Hovegaden 26, DK 2970 Horsholm, Denmark.

"Proceedings of the 4th Nondestructive Testing of Wood Symposium," Washington State University. Aug 1978.

Sherwood, G. E., "New Life for Old Dwellings," Forest Products Laboratory USDA, Handbook No. 481, December 1975.

"Wood Handbook, Wood as an Engineering Material," Forest Products Lab., USDA, Handbook No. 72, 1974.

DiGiacomo, G., Crisci, J., and Goldspiel, S.; "An Ultrasonic Method for Measuring Crack Depth in Structural Weldments," *Materials Evaluation,* September 1970.

"Fundamentals of Nondestructive Testing," by the American Society for Metals and the American Society for Nondestructive Testing—reference manuals by Metals Engineering Institute, 1979.

Halmshaw, R., editor, "Physics of Industrial Radiology," Heywood Books London, 1966.

Krautkramer, J. and Krautkramer, H., *Ultrasonic Testing of Materials,* Springer-Verlag, New York, 1969 2nd Ed.

McMaster, R. C., editor, *Nondestructive Testing Handbook,* Ronald Press, New York, 1959.

McGonnagle, W. J., *Nondestructive Testing,* Pergamon Press, New York (no date)

"Nondestructive Inspection and Quality Control," *Metals Handbook,* 8th Edition, Vol. II, American Society for Metals, 1976.

"Nondestructive Testing," reference manual for a home study and extension course, American Society for Metals, the American Society for Non-destructive Testing, and the Metals Engineering Institute, 1972.

"Radiography in Modern Industry," Eastman Kodak Co., Rochester, N.Y., 1970.

Sharp, R. S., editor, *Research Techniques in Nondestructive Testing,* Academic Press, London, 1980.

Vary, A., "Nondestructive Evaluation Technique Guide," National Aeronautics and Space Administration, Washington D.C., 1973.

METALS READINGS

AMERICAN SOCIETY FOR TESTING AND MATERIALS (ASTM)

"Standard Methods of Compression Testing of Metallic Materials at Room Temperature," ASTM E 9, 1977.

"Standard Methods of Tension Testing of Metallic Materials," ASTM E 8, 1977.

Other Readings

Dick, P., "Introduction to Nondestructive Testing," General Electric Co., for Metals Engineering Institute, 1922.

MASONRY

AMERICAN SOCIETY FOR TESTING AND MATERIALS (ASTM)

"Standard Methods of Sampling and Testing Brick and Structural Clay Tile," ASTM C 67, 1978.

"Standard Specification for Masonry Cement," ASTM C 91, 1978.

"Standard Specification for Ceramic Glazed Structural Clay Facing Tile, Facing Brick, and Solid Masonry Units," ASTM C 126, 1976.

"Standard Test Method for Air Content of Freshly Mixed Concrete by the Pressure Method," ASTM C 231, 1978.

"Standard Methods of Conducting Strength Tests of Panels for Building Construction," ASTM E 72, 1977.

"Standard Test Method for Young's Modulus at Room Temperature," ASTM E 111, 1978.

"Standard Test Methods for Compressive Strength of Masonry Prisms," ASTM E 447, 1974.

"Standard Test Methods for Water Permeance of Masonry," ASTM E 514, 1974.

"Standard Test Methods for Flexural Bond Strength of Masonry," ASTM E 518, 1976.

"Standard Test Method for Diagonal Tension (Shear) in Masonry Assemblages," ASTM E 519, 1974.

Other Readings—Masonry

"Standard Specification for Portland Cement—Lime Mortar for Brick Masonry," MI-72, Tech Note 8A, Brick Institute of America, 1972.

"The Causes and Control of Efflorescence on Brickwork," Research Report No. 15, Structural Clay Products Institute (now BIA), 1969.

Snell, L. M., "Nondestructive Testing Techniques to Evaluate Existing Masonry Construction, August 1978.

CONCRETE READINGS

AMERICAN CONCRETE INSTITUTE (ACI)

"Building Code Requirements for Reinforced Concrete," ACI 318, 1977.

Erlin, B., "Methods Used in Petrographic Studies in Concrete Paste," *Journal of the American Concrete Institute*, Vol., 27, No. 12, June 1956.

"Guide for Making a Condition Survey of Concrete in Service," ACI 201, 1975.

Leslie, J. and Cheeseman, W., "An Ultrasonic Method of Studying Deterioration and Cracking in Concrete Structures," ACI Journal, September 1949.

Mailhot, G. and Bisaillon, A.: Carette, G. G.; Malhotra, V. M.; "In Place Concrete Strength: New Pullout Methods, JOURNAL OF THE AMERICAN CONCRETE INSTITUTE, December 1979.

Malhotra, V. M., "Testing Hardened Concrete:

Nondestructive Methods," ACI Monograph No. 9, 1976.

"Practices for Evaluation of Concrete in Existing Massive Structures for Service Conditions," No. ACI 207 R-79, *Concrete International,* March 1979.

"Recommended Practice for Evaluation of Strength Test Results of Concrete," ACI 214, 1977.

"Recommended Practice for Concrete Inspection," ACI 1975.

"Strength Evaluation of Existing Concrete Buildings," ACI 1975.

Whitehurst, E. A., "Evaluation of Concrete Properties from Sonic Test," ACI Monograph No. 2, 1966.

CONCRETE

AMERICAN SOCIETY FOR TESTING AND MATERIALS (ASTM)

"Recommended Practice for Examination and Sampling of Hardened Concrete in Construction," ASTM C 823, 1975.

"Recommended Practice of Petrographic Examination of Hardened Concrete," ASTM C 856, 1977.

"Recommended Practice for Petrographic Examination of Aggregates for Concrete," ASTM C 295, 1965 rev. 1973.

"Recommended Practice for Microscopical Determination of Air-Void Content and Parameters of the Air-Void System in Hardened Concrete," ASTM C 457, 1971.

"Standard Method for Load Tests of Floors and Flat Roofs," ASTM E 196, 1974.

"Standard Method of Obtaining and Testing Drilled Cores and Sawed Beams of Concrete," ASTM C 42, 1977.

"Standard Test for Rebound Number of Hardened Concrete," ASTM C 805, 1979.

"Standard Test for Penetration Resistance of Hardened Concrete," ASTM C 803, 1975.

"Standard Test for Cement Content of Hardened Portland Cement Concrete," ASTM C 85, 1966 rev. 1973.

"Standard Test for Fineness of Portland Cement by Air Permeability Apparatus," ASTM C 204, 1979.

"Standard Test for Foundamental Transverse, Longitudinal, and Torsional Frequencies of

Concrete Specimens," ASTM C 215, 1960 rev. 1976.

"Standard Test for Static Modulus of Elasticity and Poisson's Ratio of Concrete in Compression," ASTM 469, 1965 rev. 1975.

"Standard Test for Resistance of Concrete to Rapid Freezing and Thawing," ASTM C 666, 1977.

"Standard Test for Critical Dilation of Concrete Specimens Subjected to Freezing," ASTM C 671, 1977.

"Tentative Test Method for Pullout Strength of Hardened Concrete," ASTM C 900, 1978.

NATIONAL BUREAU OF STANDARDS (NBS)

"An Infrared Technique for Measuring Heat Loss," NBSIR 78-1557, September 1977.

"Annual Report 1979," NBSIR 80-20007, 1980.

"Annual Report 1978," NBSIR 78-1581, 1978.

Clifton, J. R., "Nondestructive Tests to Determine Concrete Strength—A Status Report," NBSIR 75-729, July 1975.

"NDE (Nondestructive Evaluation)," LC 1080, April 1978.

Other Readings—Concrete Testing

Halmshaw, R., editor, *Physics of Industrial Radiology,* Heywood Books, London, 1966.

Halstead, P. E., "The Covermeter Apparatus for Measuring the Depth of Reinforcement Below the Surface of Hardened Concrete," Technical Report No. TRA 197, Cement and Concrete Association, London, July 1955.

"Nondestructive Testing Techniques," Materials Evaluation, September 1970.

"Nondestructive Testing: Trends and Techniques," Proceedings of the Second Technology Status and Trends Symposium at the Marshall Space Flight Center, October 1966.

"Radar Spots Substratum Voids Without Drilling Slabs," *Engineering News Record,* September 7, 1978.

Reiding, F. J., "A Portable Reinforcement Covermeter," Report No. BI-66-26, Institut TNO Voor, Bouchomaterialen en Bouwconstructies, Rijswijk (ZH), The Netherlands, March 1966.

Sevall, G. W., "Nondestructive Testing of Construction Materials and Operations," AD-774 847, U. S. Department of Defense, Army Construction Engineering Research Laboratory, December 1973.

U. S. Department of the Interior, "Concrete Manual—A Water Resources Technical Publication," Bureau of Reclamation, 1975 Eighth Edition.

Vary, A., "Ultrasonic Measurement of Material Properties," in "Research Techniques in Nondestructive Testing," R. S. Sharpe, editor, Academic Press, London, 1980.

Wareen, C., "Proceedings of the Third Biennial Infrared Information Exchange," The AGA Corporation, August 1976.

Bibliography

Baker, M. C., Recognition of Joints in the System: Cracks, Movements and Joints in Building, National Research Council of Canada, (1972) NRC 15477.

Baker, M. C. and Hutcheon, N. B., Cracks, Movements and Joints in Buildings, National Research Council of Canada; Record of DBR Building Science Seminar, Autumn 1972; NRCC 15477 (1976).

Botsai, Elmer, Water Infiltration in Buildings (Interview and notes from unpublished manuscript) 1981.

Brick Institute of America, Technical Notes: On Brick Construction, 1750 Old Meadow Road, McLean, Virginia 22102.

Bryant, Terry; Protecting Exterior Masonry from Water Damage, *Technology and Preservation Magazine,* Spring 1978, J.3, No.1.

Canadian Building Digest, Division of Building Research, National Research Council, Canada. ISSN 0008-3097.

Cracks, Movements and Joints in Buildings: Recordings No.2 NRCC 15477 Division of Building Research, National Research Council Canada, 1976-Record of the DBR Building Science Seminar, Autumn 1972.

Crawford, C. B., Cracks, Movements and Joints in Buildings, National Research Council of Canada; Record of DBR Building Science Seminar, Automn 1972, NRCC 15477 (1976).

Dixon, John Morris, Recipes for Baked Earth, *Progressive Architecture,* Novemeber 1977.

Feld, Jacob, Failure Lessons in Concrete Construction: A Collection of Articles from *Concrete Construction Magazine,* Addison, Illinois, 1978.

Fitzsimons, Neal C. E., Research Support for Building Rehabilitation Studies in the Area of Strength and Stability Evaluation, National Bureau of Standards, 1979.

Grimmer, Anne E., Dangers of Abrasive Cleaning to Historic Buildings: Preservation Briefs #6; Technical Preservation Services Division, Heritage Conservation and Recreation Service, U. S. Department of the Interior, Washington, D.C. 20240 (1979).

Handegord, G. O., Cracks, Movements and Joints in Buildings, National Research Council of Canada; Record of DBR Building Science Seminar, 1972; NRCC 15477 (1976).

Kelly, Joe W.; Cracks in Concrete: Causes and Prevention, A Collection of Articles from *Concrete Construction Magazine,* Addison, Illinois.

Kidd, Philip E., Value of Non-Residential Rehabilitation will Double by Mid-1980s, *Architectural Record,* page 61, October 1979.

Latta, J. K., Design of Weathertight Joints: from Cracks, Movements and Joints in Buildings; Division of Building Research, National Research Council, Canada (1972) NRCC 15477.

Latta, J. K., Cracks, Movements and Joints in Buildings: National Council of Canada; Record of DBR Building Science Seminar, Autumn 1972, NRCC 15477 (1976).

Lerchen, Pielert, Faison, Selected Methods for Condition Assessment of Structural, HAVAC, Plumbing and Electrical Systems in Existing Buildings; U. S. Dept. of Commerce, National Bureau of Standards, Washington D.C. (1980).

Mack, Robert C., The Cleaning and Waterproof Coating of Masonry Buildings: Preservation Briefs #1, Technical Preservation Services Division, Office Archeology and Historic Preservation/Heritage Concervation and Recreation Service; U. S. Department of the Interior, Washington D.C. 20240 (1975).

Mack, Robert C., Tiller, de Teel Patterson, Askins, James, Repointing Mortar Joints in Historic Brick Buildings: Preservation Briefs #2, Heritage Conservation and Recreation Service, Technical Preservation Services Division; U. S. Department of the Interior, Washington D.C. 20243.

Metals Handbook; American Society for Metals, Volume 11, Nondestructive Inspection and Quality Control, Metals Park, Ohio 44073.

Nelson, Lee H., Preservation Services Division, Office of Archeology and Historic Preservation/Heritage Conservation and Recreation Service.

Olin, Harold, Schmidt, John, and Lewis, Walter; Construction: Principles, Materials and Methods; The Institute of Financial Education, Chicago, Illinois and Interstate Printers and Publishers, Danville, Illinois (1980) (Fourth Edition).

Parker, Gay, McGuire; Materials and Methods of Architectural Construction (3rd Edition) John Wiley & Sons, New York (1958).

Plewes, W. G., Cracks, Movements and Joints in Buildings; National Research Council of Canada; Record of DBR Building Science Seminar, Autumn 1972, NRCC 15477 (1976).

Richie, T, Cleaning of the Brickwork; Canadian Building Digest Division of Building Research, National Research Council of Canada, April 1978 CBD 194.

Sweetser, Sarah M., Roofing for Historic Buildings: Preservation Briefs #4, Technical Preservation Services Division, Office of Archeology and Historic Preservation, Heritage Conservation and Recreation Service; U. S. Department of the Interior, Washington, D.C. 20240 (1978).

Tenzer, A. J. and McNeal, Ray, Building Materials Research Institute N.Y.C., *Progressive Architecture,* October 1966 (p. 221).

Tiller, de Teel Patterson, et al, Preservation of Historic Adobe Buildings; Preservation Briefs #5; Technical Preservation Services Division, Office of Archeology and Historic Preservation/Heritage Conservation and Recreation Service; Department of the Interior, Washington D.C. 20240 (1978).

Tiller, de Teel Patterson, The Preservation of Historic Glazed Architectural Terra-Cotta: Preservation Briefs #7; Technical Preservation Services Division, Heritage Conservation and Recreation Service, U. S. Department of the Interior, Washington, D.C. 20240 (1979).

Weiss, Norman R., Exterior Cleaning of Historic Masonry Buildings: (Draft) of paper for Heritage Conservation and Recreation Service, Office of Archeology and Historic Preservation.

Wood Handbook, U. S. Dept. of Agriculture, Forest Products Laboratory; Forest Service Agriculture Handbook No.74, U. S. Goverment Printing Office, Washington D.C. (1974).

U. S. Dept. of Commerce Construction Reports, Residential Alterations and Repairs, Annual 1979, issued April 1980.

USDA Forest Service, Research Paper FPL 190 (1973) U. S. Department of Agriculture, Forest Service, Forest Products Laboratory, Madison, Wisconsin.

Yoder, B; Dadson E. J.; Nondestructive Testing in Building Construction, *Progressive Architecture,* September 1966.

Index

compression load cracks, 224
concrete, 205
  aggregates, 215
  failure, 6
Concrete Construction Magazine, 216, 217
concrete cracking, 42
  causes and cures, 216, 218
concrete masonry strength, 122
concrete reinforcement, 55
concrete test methods-chart, 232-237
condensation, 61, 63, 103
  wood, 157
Construction: Principles, Materials and Methods, by
    Harold B. Olin, John L. Schmidt, Walter H.
    Lewis; The institute of Financial Education and
    Interstate Printers, 14, 16, 35, 61-65, 101, 125,
    188
consolidation settlement, 52
contraction joints, concrete 218
cooked earth (adobe), 90
core drilling, concrete, 253, 254
cornice and overhang problems, 134
corrosion, 34, 35
corrosion resistance, 183
  aluminum 190
coupon test, metals, 202
cover meters, concrete, 244
covalent bonding, 20
cracking, factors affecting concrete, 218
cracking process, concrete, 217
cracking restraint, 41
cracking symbols, 295
cracking under load, concrete, 224
cracks, concrete masonry, 124
Crawford, C.B., 51-54
crawl space, 157, 159, 163
  water vapor, 64
crazing, cracks, 225
creep, 43
  concrete, 51
creep elongation, 45
crystallization, masonry decay, 129
curing
  concrete, 219
  concrete masonry, 124
curtain walls, 132
cyclical movements, concrete, 239

damage, wood, 160
damping capacity, 32
damp-proofing, 101
decay, wood, 156, 159, 161, 162, 167
  extent of 169
decay resistance, wood, 176-178
defects and decay testing, wood, 172, 173
deformation, 40, 49, 51
  and potential movement, 55
  foundation, 51-53
degree of wood degradation, 169, 170
dehumidification, 61

delectric measurements, concrete, 242-243
density testing, wood, 172, 173
destructive and non-destructive testing, 9
destructive evaluation, metals, 201
destructive testing, 9
destructive tests, concrete, 253
diagnosis, 13
  medical, 13
dielectric-type moisture meter, wood, 170
differential settlement, 53, 54
dimensional change, 46-49
dirt, definition, 269-270
discharge test, drainage and vent systems, 314
discoloration, wood, 161
distintegration, concrete, 215, 216
Dixon, John Morris, 90, 93, 99
dormant cracks, concrete, 225
drying shrinkage, concrete masonry, 124
ductility, 23, 27, 28, 29
dusting, concrete, 225
dynamic tests, ultrasonic pulse velocity and resonant
    frequency, concrete, 247-248

eddy-current inspection, 330
eddy current, metals, 198, 199
Eddystone Lighthouse, 211
Edison, Thomas, 212
efflorescence, 106, 108-112
  concrete, 225
elastic deformation, 23
elastic limit, 25, 26, 27
electric branch circuits test, 316-317
electric co-heating, 321-322
electrical conduit, concrete, 216
electrical impedence hygrometer, 321
electrical properties, 34
electrical resistance probe, wood, 172
electrical resistivity test, concrete, 243
electrical tests, concrete, 242
electrolyte, 34
environmental concerns, cleaning, 271
equilibrium moisture content, wood, 150
efflorescence, 129
evaluation, masonry, 121
evaporation, concrete, 219, 221
exposure, concrete, 222
extruding, 187

Faison, Thomas K., 10, 121, 167, 196, 237
Fathy, Hassan, 83, 84, 88
fatigue, 31
Feld, Jacob, 6, 9, 215, 216
ferrous oxide, cement, 130
fiber optics, concrete, 240
fiber saturation, wood, 150
Fitzsimons, Neal, 10, 122, 138, 281-298
flashing, 107
form scabbing, concrete, 225
freeze-thaw cycles, 115
  concrete, 210

metamorphic rocks, 136
methods of working, metal, 186
micro cracks, stone, 126
microwave absorption, concrete, 247
mistakes multiply, precasting, 262
modulus of elasticity, 23, 25, 43, 45
modulus of elasticity testing, wood, 172, 173
molecular materials, 17, 22
molecules, 15
mold, 160
moisture
  and chemical reactions, 46
  dimensional change, 46
  masonry, 101
  penetration, brick, 102
moisture content
  in concrete masonry, 124
  in wood, 150, 170, 171, 172, 174, 175, 176
mortar, 96
  properties, 114-116
  strength, 115
  uneven movement, 115
movement, building materials, 55-58
movements, potential, 54
mud brick, 81-88
  stabilized, 82

National Research Council of Canada, 215
National Bureau of Standards (NBS), 71
natural cement, 211
Nelson, Lee, 115, 116
neutron activation analysis, 249
neutron moisture gauges, concrete, 248-249
neutron radiography, 332
New York City, deterioration, 1
non-destructive evaluation, 9
  of metals, 193
non-destructive tests, concrete, 230
nonporous materials, 47

Olin, Harold B., Schmidt, John L., Lewis, Walter
  H., 14, 35, 61-65, 101, 125, 188, 191
oven-dry test, wood, 147, 174, 175

painted surfaces, visual inspection, 278-280
parapet problems, 134
patching and repairing, adobe, 87
performance requirements, joints, 302-304
permeability, concrete, 226
permeance, 63
perms, 63
petrography, concrete, 254-256
physical inspection
  of furnaces and boilers, 322-323
  of plumbing components, 311-312
pick test, 162
Pielert, James H., 10, 121, 167, 196, 237
pig iron, 182
pilodyn, wood tester, 169
pitch pockets, 145, 154, 155

placing, concrete, 219
plastic cracking, concrete, 219
plastic deformation, 23
plastics, 22
plastics flashing, 107, 108
Plewes, W.G., 46-51
plumbing systems, test methods, 311
polyethylene pipe, 45
polyvinyl chloride, 45
popouts, concrete, 226
porosity, 103
  and permeability, stone, 128-129
porous building materials, 47
Portland cement, 6
Portland Cement Association, 212
portland cement mortars, 133
portland limestone and brick problems, 130
precast concrete panels, 259-266
preliminary steps, structural assessment, 291-292
pressure differential, 67-73
pressure differential plane, theory, 70
prestressed concrete, cracking, 224
principles of wall design, 301-302
problems checklist, concrete, 256-259
procedures for building assessment, 281
Progressive Architecture, 99, 211, 266
project dossier, structural assessment, 292-293
properties, wood, 147
protective coatings and mechanical finishes, 188
pullout text, concrete, 251-252
pulse velocity equipment-wood, 172
Prudon, Theodore, H.M., 93

radar test, concrete, 251
radiographic evaluation, concrete, 246-247
radiographic evaluation (X-ray), metals, 197-198
radiographic inspection, 330-332
radiography testing, wood, 175-176
rain penetration, 74, 103, 104
Ransome, Earnest, 212
rapid identification, metals and alloys, 334-335
rebars, 212
rebound test, concrete, 241
rehabilitation project model, 297
reinforced concrete, history, 212
reinforcing steel, 6
relative humidity, 47, 48
  in building materials, 103
repointing, 113, 114
residential case study, 336-342
residual strain in structures, 192
resistance-type moisture meter, wood, 171, 172
restraint, concrete, 222
ribbed panels, 76
Richie, T., 113
rising damp, 104, 114
Robertson, Eugene C., 128, 138
Robinson, Gilbert C., 138
roofs, 71
roof drainage, 134